W9-AAR-989

The ESSENTIALS of

REGISTERED TRADEMARK

ORGANIC CHEMISTRY I

Staff of Research and Education Association,
Dr. M. Fogiel, Director

This book covers the usual course outline of Organic Chemistry I. For more advanced topics, see *"THE ESSENTIALS OF ORGANIC CHEMISTRY II"*.

Research and Education Association
61 Ethel Road West
Piscataway, New Jersey 08854

THE ESSENTIALS ®
OF ORGANIC CHEMISTRY I

1997 PRINTING

Printed in the United States of America

Library of Congress Catalog Card Number 94-69635

International Standard Book Number 0-87891-616-4

ESSENTIALS is a registered trademark of Research & Education Association, Piscataway, New Jersey 08854

WHAT "THE ESSENTIALS" WILL DO FOR YOU

This book is a review and study guide. It is comprehensive and it is concise.

It helps in preparing for exams, in doing homework, and remains a handy reference source at all times.

It condenses the vast amount of detail characteristic of the subject matter and summarizes the **essentials** of the field.

It will thus save hours of study and preparation time.

The book provides quick access to the important facts, principles, theorems, concepts, and equations of the field.

Materials needed for exams can be reviewed in summary form — eliminating the need to read and re-read many pages of textbook and class notes. The summaries will even tend to bring detail to mind that had been previously read or noted.

This "ESSENTIALS" book has been carefully prepared by educators and professionals and was subsequently reviewed by another group of editors to assure accuracy and maximum usefulness.

Dr. Max Fogiel
Program Director

CONTENTS

ix

CHAPTER 1

STRUCTURE AND PROPERTIES

1.1 ATOMIC AND MOLECULAR ORBITALS

Atomic orbitals are arrangements of electrons around the nucleus of an atom. An electron occupies an orbital according to its energy content. In order of increasing energy, the orbitals are specified by the letters, s, p, d, and f, within a given shell. The shells are also arranged in order of increasing energy and are assigned the letters K, L, M, etc.

In Fig. 1-1, the shapes of some of these orbitals are shown.

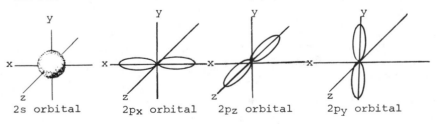

2s orbital   2p$_x$ orbital   2p$_z$ orbital   2p$_y$ orbital

Fig. 1-1 Atomic Orbitals (s and p)

The overlapping of atomic orbitals leads to the formation of molecular orbitals, and thus molecular bonding (covalent).

1

The sigma (σ) bond, with its characteristic shape, is formed from the overlapping of two s-orbitals, two p-orbitals, or an s and a p-orbital.

Two molecular orbitals, one bonding and one antibonding, are formed when two atomic orbitals are joined. The bonding orbital is of lower energy and is more stable than the component atomic orbitals. The antibonding orbital is of higher energy and is less stable than the component atomic orbitals. This is shown in Fig. 1-2.

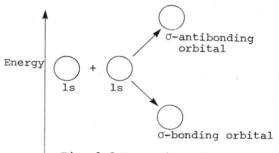

Fig. 1-2 Formation of two molecular orbitals

Electrons in antibonding orbitals lead to repulsive forces, which are almost as strong as the attractive forces in bonding orbitals.

1.2 ELECTRON CONFIGURATION

The Pauli exclusion principle states that only two electrons can occupy an atomic orbital, and these two must have opposite spins. Electrons with like spins cannot occupy the same orbital.

Table 1-1 shows the electronic configurations for the first ten elements of the periodic table.

Following the examples shown in the table, the electronic configuration of Argon is expressed as

$$1s^2\, 2s^2\, 2p_x^2\, 2p_y^2\, 2p_z^2\, 3s^2\, 3p_x^2\, 3p_y^2\, 3p_z^2$$

2

or equivalently

$$1s^2 2s^2 2p^6 3s^2 3p^6.$$

Table 1-1 Electronic Configurations

Symbol	Atomic Number		1s	2s	$2p_x$	$2p_y$	$2p_z$
H	1	$1s$	↑				
He	2	$1s^2$	↑↓				
Li	3	$1s^2 2s$	↑↓	↑			
Be	4	$1s^2 2s^2$	↑↓	↑↓			
B	5	$1s^2 2s^2 2p_x$	↑↓	↑↓	↑		
C	6	$1s^2 2s^2 2p_x 2p_y$	↑↓	↑↓	↑	↑	
N	7	$1s^2 2s^2 2p_x 2p_y 2p_z$	↑↓	↑↓	↑	↑	↑
O	8	$1s^2 2s^2 2p_x^2 2p_y 2p_z$	↑↓	↑↓	↑↓	↑	↑
F	9	$1s^2 2s^2 2p_x^2 2p_y^2 2p_z$	↑↓	↑↓	↑↓	↑↓	↑
Ne	10	$1s^2 2s^2 2p_x^2 2p_y^2 2p_z^2$	↑↓	↑↓	↑↓	↑↓	↑↓

1.3 HYBRID ORBITALS

The sp hybrid orbitals arise from the mixing of one s orbital and one p orbital. These orbitals are equivalent and much more strongly directed than either the s or p orbital. The sp hybrid orbitals point in exactly opposite directions, which permits them to get as far away from each other as possible.

The sp^2 hybrid orbitals arise from the mixing of one s orbital and two p orbitals. These orbitals lie in a plane which includes the atomic nucleus. They are directed to the corners of an equilateral triangle with an angle of 120° between any of two orbitals.

The sp^3 hybrid orbitals arise from the mixing of one s orbital and three p orbitals. These orbitals are directed to the corners of a regular tetrahedron. The angle between any two orbitals is the tetrahedral angle 109.5°.

1.4 CHEMICAL BONDING

An ionic bond is the electrostatic attraction between oppositely charged particles, which results from the transfer of electrons.

An ion-dipole bond is formed if one of the ions in an ionic bond is replaced by a highly polar molecule, such as water. This bond results from the attraction of the ion to the oppositely charged end of the polar molecule.

A dipole-dipole bond is formed if the ion in an ion-dipole bond is replaced with another polar molecule. This bond results from the attraction of the oppositely charged ends of the two polar molecules.

The ionic bonds form stronger bonds than ion-dipole bonds which in turn are stronger than dipole-dipole bonds.

The formation of covalent bonds by the sharing of electrons results from the overlapping and interaction of partially filled atomic orbitals.

The degree of the overlapping of the atomic orbitals to form a bonding molecular orbital determines the strength of the covalent bonds.

Bond length is the distance between bonded nuclei. At this distance the repulsion that occurs between the similarly-charged nuclei balances the "packing" effect of bonding.

Sigma (σ) bonds occur from the formation of σ-orbitals, which result from the end-to-end overlap of s and p atomic orbitals. It is also possible for the p orbitals to overlap side-to-side, in which case they from Pi (π) bonds. These bonds consist of π orbitals above and below the plane of the molecule. These are always present in double and triple bonds.

1.5 BOND DISSOCIATION ENERGY

The amount of energy liberated when a bond is formed is called the bond dissociation energy. This is also the energy needed to break the bond. The more energy that is lost when a bond is formed the stabler the bond will be and the more energy will have to be used in breaking, or dissociating, it.

The breakage of a covalent bond resulting in an equal distribution of electrons to each of the fragments is known as **homolysis**.

$$A : B \;\rightarrow\; A\cdot + B\cdot$$

An unequal distribution of electrons is known as heterolysis.

$$A:B \;\rightarrow\; A: + B \text{ or } A + B:$$

Simple heterolysis of a neutral molecule yields a positive ion and a negative ion. In the gas phase, bond dissociation generally takes place by homolysis. In an ionizing solvent, heterolysis occurs more frequently.

1.6 STRUCTURE AND PHYSICAL PROPERTIES

POLARITY OF BONDS

Polarity results from an unequal distribution of the electron cloud about two nuclei, such that the negative pole exists where the electron cloud is denser about one nucleus, and the positive pole exists where the electron cloud is sparser about the other nucleus.

A covalent bond will be polar if it joins atoms that differ in their tendency to attract electrons, that is, atoms that differ in electronegativity. The greater the difference in electronegativity, the more polar the bond will be. The following relates the electronegativity of some common elements: F > O > Cl, N >Br > C > H.

Examples of polar bonds are:

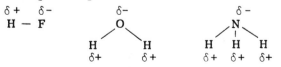

The symbols $\delta+$ and $\delta-$ indicate partial charges and are used to indicate polarity.

POLARITY OF MOLECULES

When the center of the negative charge of a molecule does not agree with the center of the positive charge, the molecule is polar.

A dipole is defined as a molecule with two equal and opposite charges separated in space. The dipole possesses a dipole moment, μ, which is equal to the magnitude of the charge, e, multiplied by the distance, d, between the centers of the charge:

$$
\begin{array}{ccc}
 & \mu = e \times d & \\
\text{in DEBYE} & \text{in} & \text{in} \\
\text{units, D} & \text{e.s.u.} & \text{cm}
\end{array}
$$

The values of the dipole moments indicate the relative polarities of different molecules. The following are dipole moments for some molecules:

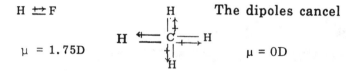

H $\leftrightarrows$ F

$\mu = 1.75D$

The dipoles cancel

$\mu = 0D$

The symbol $\longmapsto$ represents a dipole, where the arrow points from positive to negative. Molecules, such as that of methane, with zero dipole moments are nonpolar.

MELTING POINT

Melting occurs at the temperature where the kinetic energy of the particles is great enough to overcome the forces binding the particles. Melting is defined as a change from an ordered arrangement of particles in a crystalline lattice to a more random arrangement in a liquid.

An ionic compound forms crystals whose structural units are ions. Only at very high temperatures can the strong interionic forces be overcome.

In a non-ionic compound, the atoms are held together by covalent bonds and form crystals in which the structural units are molecules. Non-ionic compounds melt at lower temperatures than ionic compounds due to weak intermolecular forces.

INTERMOLECULAR FORCES

Dipole-dipole interaction is the attraction of the positive end of one polar molecule for the negative end of another polar molecule. Polar molecules are held more strongly together than are non-polar molecules of comparable weight.

An example of a strong dipole-dipole interaction is hydrogen bonding, in which a hydrogen atom serves as a bridge between two electronegative atoms, holding one by a covalent bond and the other by purely electrostatic forces. The hydrogen bond is indicated by the broken line as follows: H — F --- H— F.

For hydrogen bonding to be important, the hydrogen atom must be attached to a highly electronegative atom, such as fluorine, oxygen, or nitrogen.

The forces between the molecules of a non-polar compound are called van der Waals forces. These forces result from the formation of small momentary dipoles that are created from the instantaneous asymmetrical distribution of electrons and the attraction of these to similar momentary dipoles. These forces have a very short range in that they act only between portions of different molecules in close touch. Every atom has an effective "size" called its van der Waals radius.

BOILING POINT

Boiling occurs at the temperature where the kinetic energy of the particles is great enough to overcome the cohesive forces that hold them in the liquid state.

The ion is the basic unit of an ionic compound in the liquid state. In order for a pair of oppositely charged ions to break away from the liquid, a great amount of energy is required. For this reason boiling occurs at very high temperatures.

The molecule is the basic unit of a non-ionic compound in the liquid state. Boiling for non-ionic compounds occurs at much lower temperatures than for ionic compounds because of the relatively weak intermolecular forces that must be overcome.

Associated liquids are liquids with molecules held together by hydrogen bonds. Because of the strength of these bonds, these liquids also boil at high temperatures.

Since molecular size is proportional to the strength of the van der Waals forces, polarity, hydrogen bonding, and thus boiling points, an increase in the size of the molecule corresponds to an increase in these properties.

SOLUBILITY

In order for an ionic compound to dissolve, the solvent must be able to form ion-dipole bonds and have a high dielectric constant. Only water or other highly polar solvents are able to dissolve ionic compounds appreciably.

Solubility of non-ionic compounds is determined by their polarity. Non-polar or weakly polar compounds dissolve in non-polar or weakly polar solvents. Highly polar compounds dissolve in highly polar solvents. Non-ionic compounds thus follow the rule, "like dissolves like."

Water is a poor solvent for most organic compounds. Solvents such as water and methanol are examples of protic solvents. These are acidic solvents containing hydrogen atoms attached to oxygen or nitrogen atoms.

1.7 ACIDS AND BASES

The Lowry-Bronsted definition states that an acid is a substance that donates a proton, and a base is a substance that accepts a proton.

The strength of an acid depends upon its tendency to donate a proton, and the strength of a base depends upon its tendency to accept a proton. The relative strengths of some common acids and bases:

$$\text{Acid strength} \quad \begin{matrix} H_2SO_4 \\ HCl \end{matrix} > H_3O^+ > NH_4^+ > H_2O$$

$$\text{Base strength} \quad \begin{matrix} HSO_4^- \\ Cl^- \end{matrix} < H_2O < NH_3 < OH^-$$

The Lewis definition states that a base is a substance that can donate an electron pair to form a covalent bond, and an acid is a substance that can accept an electron pair to form a covalent bond. Thus, an acid is an electron-pair acceptor and a base is an electron-pair donor.

The ability to accommodate the electron pair depends upon several factors, including:

a) the atom's electronegativity, and

b) its size.

Within a given row of the periodic table, acidity increases as electronegativity increases:

Acidity $H-CH_3 < H-NH_2 < H-OH < H-F$

$H-SH < H-Cl$

within a given family, acidity increases as the size increases:

Acidity $H-F < H-Cl < H-Br < H-I$

$H-OH < H-SH < H-SeH$

CHAPTER 2

ALKANES

Structural Formula: C_nH_{2n+2}

The simplest member of the alkane family is methane (CH_4) which is written as:

$$
\begin{array}{ccc}
\overset{\displaystyle H}{\underset{\displaystyle H}{H-C-H}} & \text{or} & H:\overset{\displaystyle H}{\underset{\displaystyle H}{C}}:H
\end{array}
$$

2.1 NOMENCLATURE (IUPAC SYSTEM)

A) Select the longest continuous carbon chain for the parent name.

Ex. $CH_3-CH_2-CH-CH_2-CH_2-CH_3$
              |
              CH_3

The parent name is hexane.

B) Number the carbons in the chain, from either end, such that the substituents are given the lowest numbers possible.

Ex. $\overset{1}{C}H_3-\overset{2}{C}H_2-\overset{3}{C}H-\overset{4}{C}H_2-\overset{5}{C}H_2-\overset{6}{C}H_3$
                          |
                          CH_3

C) The substituents are assigned the number of the carbon to which they are attached. In the preceding example the substitutent is assigned the number 3.

D) The name of the compound is now composed of the name of the parent chain preceded by the name and the number of the substituents, arranged in alphabetic order. For the same example the name is thus 3-methylhexane.

E) If a substituent occurs more than once in the molecule, the prefixes, "di-," "tri-," "tetra-," etc., are used to indicate how many times it occurs.

F) If a substituent occurs twice on the same carbon, the number of the substituent is repeated.

Ex.

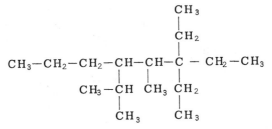

3,3-diethyl-5-isopropyl-4-methyloctane

2.2 PHYSICAL PROPERTIES OF ALKANES

Alkanes consist of non-polar or very weakly polar molecules, which are attracted to each other in the liquid or solid phase by weak van der Waals forces.

The boiling point, melting point, density and viscosity increase as the length of the carbon chain increases.

In general, branched chain alkanes exhibit lower boiling points than corresponding straight-chain alkanes.

The alkanes are soluble in all non-polar or weakly polar solvents, such as benzene, chloroform, ether and carbon tetrachloride, because "like dissolves like."

The melting point of an alkane is not only dependent on the size of the molecule, but also on the ease with which the molecule fits into a crystal lattice.

2.3 ROTATIONAL STRUCTURES

The atoms in alkane molecules are joined by σ-bonds, and the electron distribution is cylindrically symmetrical about a line joining the atomic nuclei.

There exists an infinite number of free rotations about the carbon-carbon single bonds of the alkanes. Each rotation results in the rearrangement of the atoms of the molecules and is called a conformation.

Take for example ethane, with its three-dimensional representation shown in Fig. 1. The frontal carbon atom is represented by a point and the atom behind it, by a circle; such representation is called a Newman projection.

If the frontal carbon atom in (a) is rotated 60° and the other is held fixed, a second Newman projection is achieved, as shown in (b). These two arrangements of the atoms of the molecule are referred to respectively as the staggered and eclipsed conformations of ethane.

a   b

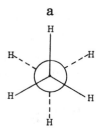   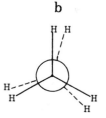

a) staggered conformation   b) eclipsed conformation
Fig. 1. Newman Projections of Ethane

2.4 PREPARATION

HYDROGENATION OF ALKENES

Alkenes are converted into alkanes of the same carbon skeleton by the addition of hydrogen atoms to the double bond.

$$C_nH_{2n} \xrightarrow[\text{or Ni}]{H_2 + Pt} C_nH_{2n+2}$$

HYDROLYSIS OF GRIGNARD REAGENTS

Grignard reagents are organometallic compounds, RMgX (i.e. compounds that contain carbon-metal covalent bonds). They are very reactive and can be prepared by reaction of magnesium metal, Mg, and an alkyl halide, R-x, in dry ether solvent.

$$R-X + Mg \xrightarrow{\text{dry ether}} RMgX \xrightarrow{H_2O} R-H$$

alkyl halide   Grignard   alkane
                  reagent

where: X = halide (Cl, Br or I)

Example:

$$\underset{\text{isobutyl bromide}}{CH_3-\underset{\underset{CH_3}{|}}{CH}-CH_2-Br} \xrightarrow[\text{ether}]{Mg} CH_3-\underset{\underset{CH_3}{|}}{CH}-CH_2-MgBr$$

$$\xrightarrow[-Mg(OH)Br]{+H_2O} \underset{\text{isobutane}}{CH_3-\underset{\underset{CH_3}{|}}{CH}-CH_3}$$

REDUCTION BY METAL AND ACID

Alkyl halides can be reduced to alkanes by reaction with zinc metal and a mineral acid.

$$R-X + Zn + H^{\oplus} \rightarrow R-H + Zn^{2\oplus} + X^{\ominus}$$

Example:

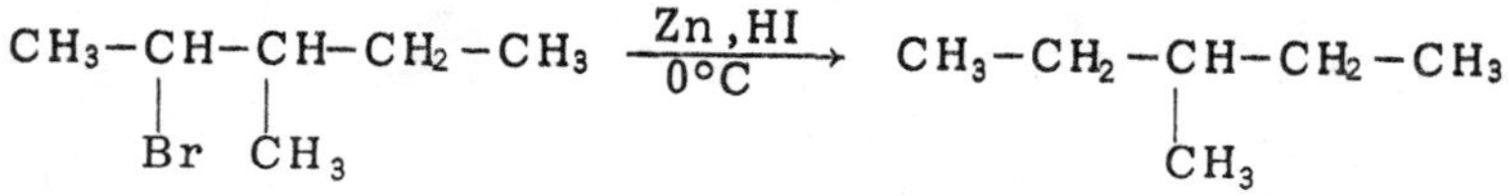

$$CH_3-CH-CH-CH_2-CH_3 \xrightarrow[0°C]{Zn,HI} CH_3-CH_2-CH-CH_2-CH_3$$

with substituents Br and CH_3 on left structure, CH_3 on right structure

2-bromo-3-methylpentane   3-methylpentane

REDUCTION WITH ALKALI METAL HYDRIDES

The strong reducing agents, lithium aluminum hydride, $LiAlH_4$, and sodium borohydride, $NaBH_4$, readily reduce alkyl halides to alkanes.

$$\boxed{R-X + LiAlH_4 \xrightarrow{\text{dry ether}} R-H}$$

Examples:

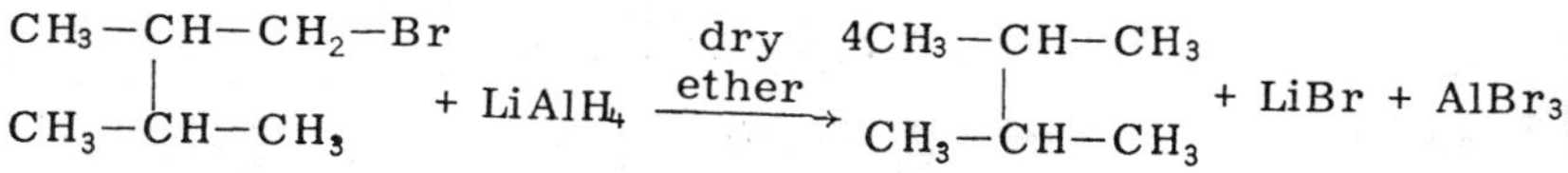

$$CH_3-CH-CH_2-Br$$
$$CH_3-CH-CH_3 + LiAlH_4 \xrightarrow[\text{ether}]{\text{dry}} \begin{array}{c} 4CH_3-CH-CH_3 \\ CH_3-CH-CH_3 \end{array} + LiBr + AlBr_3$$

1-Bromo-2, 3-dimethylbutane   2,3-dimethylbutane

THE WURTZ REACTION

When 2 moles of an alkyl halide and 2 moles of metallic sodium are reacted, an alkyl sodium is formed which reacts with a second alkyl halide to form the alkane:

$$R-X + 2Na \longrightarrow R-Na + NaX$$

$$R-Na + X-R \longrightarrow R-R + NaX$$

_____

$$2R-X + 2Na \longrightarrow R-R + 2NaX$$

The reaction between sodium and two different alkyl halides yields a mixture of three different alkanes.

$$3RX + 3R'X + 6Na \rightarrow R-R + R'-R' + R'-R + 6NaX$$

THE KOLBE SYNTHESIS

The electrolysis of sodium, potassium or calcium salts of carboxylic acids yields alkanes.

$$2CH_3-CH_2-CO\overset{\ominus}{O} \ Na^{\oplus} + 2H_2O \xrightarrow{\text{electrolysis}} CH_3-CH_2-CH_2-CH_3$$

sodium propanoate   n-butane (on anode)

$$2CO_2 + 2NaOH + H_2$$

on cathode

2.5 REACTIONS OF ALKANES

OXIDATION OF ALKANES

The combustion of alkanes gives carbon dioxide and water, with the evolution of large quantities of heat.

$$CH_4 + 2O_2 \xrightarrow{\text{flame}} CO_2 + 2H_2O$$

$$2 CH_3CH_3 + 7O_2 \xrightarrow{\text{flame}} 4CO_2 + 6H_2O$$

FORMATION OF ALKYL HYDROPEROXIDES

Good yields are obtained from preparation with tertiary carbons.

Example:

$$\underset{\text{isobutane}}{CH_3-\underset{\underset{CH_3}{|}}{\overset{\overset{CH_3}{|}}{C}}-H} + O_2 \xrightarrow{140°C} \underset{\text{tert-butylhydroperoxide}}{CH_3-\underset{\underset{CH_3}{|}}{\overset{\overset{CH_3}{|}}{C}}-O-O-H}$$

HALOGENATION OF ALKANES

In the presence of heat or ultraviolet light, either chlorine or bromine reacts with an alkane to produce an alkyl halide and a hydrohalic acid.

15

Example: Chlorination of Methane

$$\underset{\text{methane}}{H-\overset{\displaystyle H}{\underset{\displaystyle H}{C}}-H} \; + \; Cl-Cl \; \xrightarrow[\text{or peroxides}]{\text{light,} \;\; \text{heat}} \; \underset{\text{chloromethane}}{H-\overset{\displaystyle H}{\underset{\displaystyle H}{C}}-Cl} \; + \; H-Cl$$

This is an example of a substitution reaction. On further halogenation, all hydrogen atoms in the methane molecule will be replaced by halogen atoms.

The reactivity of a hydrogen atom during the halogenation process depends chiefly on its class and not on the alkane to which it is attached. The relative ease with which hydrogen atoms can be abstracted is:

$$3° > 2° > 1° > CH_4$$

2.6 STRUCTURAL ISOMERISM

There are 2 types of butanes — normal butane and iso-butane. They have the same molecular formula, C_4H_{10}, but have different structures. n-Butane is a straight chain hydrocarbon whereas iso-butane is a branched-chain hydrocarbon. n-Butane and iso-butane are structural isomers and differ in their physical and chemical properties.

$C_4H_{10} \equiv CH_3CH_2CH_2CH_3 \equiv$

n-butane
(straight chained)

$C_4H_{10} \equiv CH_3CH(CH_3)CH_3 \equiv$

isobutane
(branched)

16

In higher homologs of the alkane family, the number of isomers increases exponentially.

2.7 FREE RADICAL REACTIONS

CHLORINATION BY SULFURYL CHLORIDE

$$R-H + SO_2Cl_2 \xrightarrow[40-80^\circ C]{\text{light or peroxide}} R-Cl + SO_2 + HCl$$

SULFOCHLORINATION

Alkanes are sulfochlorinated by sulfuryl chloride, in the presence of a base to form alkane sulfonylchlorides.

$$R-H + SO_2Cl_2 \xrightarrow{\text{base}} R-SO_2Cl + HCl$$

NITRATION

In the vapor phase, alkanes can be nitrated, at high temperature, with HNO_3 or N_2O_4 to form a mixture of nitroalkanes.

$$R-H + HNO_3 \xrightarrow{> 400^\circ C} R-NO_2 + H_2O$$

CHAPTER 3

ALKENES

Alkenes (olefins) are unsaturated hydrocarbons with one or more carbon-carbon double bonds. They have the general formula, C_nH_{2n+2}.

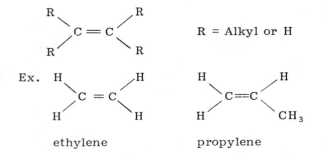

R = Alkyl or H

Ex.

ethylene   propylene

3.1 NOMENCLATURE (IUPAC SYSTEM)

A) Select the longest continuous chain of carbons containing the double bond. This is the parent structure and is assigned the name of the corresponding alkane with the suffix changed from "-ane" to "-ene."

B) Number the chain so that the position of the double bond is designated by the lowest possible number assigned to the first doubly bonded carbon.

Ex.
$$\overset{5}{C}H_3-\overset{4}{C}H_2-\overset{3}{C}H=\overset{2}{C}H-\overset{1}{C}H_3$$
$$|$$
$$Br$$

4-bromo-2-pentene

Some common names given to families of alkenes are:

$H_2C = CH-R$ vinyl

$H_2C = CH-CH_2-R$ allyl

$H_3C-CH = CH-R$ propenyl

3.2 PHYSICAL PROPERTIES OF ALKENES

A) Alkenes have lower densities than that of water.

B) C_1 to C_4 are gases.

C) C_5 to C_{15} are liquids.

D) $> C_{16}$ are solids.

E) The boiling point, melting point, viscosity, and specific gravity increase with an increase in the length of the carbon chain.

F) Branching lowers the boiling point of alkenes.

G) Alkenes are relatively insoluble in water, but are soluble in non-polar solvents such as benzene, ether, and chloroform.

H) Alkenes are colorless.

I) Alkenes show relatively higher reactivity than alkanes.

3.3 PREPARATION OF ALKENES

A) Dehydrohalogenation of Alkyl Halides

$$-\overset{\displaystyle |}{\underset{\displaystyle |}{C}}-\overset{\displaystyle |}{\underset{\displaystyle |}{H}}- + KOH \xrightarrow{\text{Alc.}} -\overset{\displaystyle |}{C}=\overset{\displaystyle |}{C}- + KX + H_2O$$

H X   strong
        base

Ex. $CH_3CH_2CH_2Br \xrightarrow[\text{alc.}]{KOH} CH_3CH = CH_2 + KBr + H_2O$

1-bromopropane   propylene

Ease of dehydrohalogenation of alkyl halides is: $3° > 2° > 1°$

B) Dehalogenation of Vicinal Dihalides

$$-\overset{\displaystyle |}{\underset{\displaystyle |}{C}}-\overset{\displaystyle |}{\underset{\displaystyle |}{C}}- + Zn \rightarrow -\overset{\displaystyle |}{C}=\overset{\displaystyle |}{C}- + ZnX_2$$

X X

Ex. $CH_3CH-CH_2 + Zn \rightarrow CH_3CH=CH_2 + ZnBr_2$
$\qquad\quad$ |$\quad$ |$\qquad\qquad\qquad$ propylene
$\qquad\quad$ Br$\;$ Br

1,2-dibromopropane

C) Dehydration of Alcohols

$$-\overset{\displaystyle |}{\underset{\displaystyle |}{C}}-\overset{\displaystyle |}{\underset{\displaystyle |}{C}}- \xrightarrow[\text{Heat}]{\text{Acid}} -\overset{\displaystyle |}{C}=\overset{\displaystyle |}{C}- + H_2O$$

H OH

Ex. $CH_3CH_2OH \xrightarrow[170°C]{H_2SO_4} CH_2 = CH_2$

ethanol   ethylene

Ease of dehydration of alcohols is: $3° > 2° > 1°$.

D) Pyrolysis (Cleavage by Heat)

Alkane $\xrightarrow[\text{without catalyst}]{\text{400-600°C; with or}}$ smaller alkane + alkene + H_2

Ex. $CH_3CH_2CH_2CH_2CH_2CH_3 \rightarrow CH_3CH_2CH_2CH_3 + CH_2 = CH_2 + H_2$

hexane   butane   ethylene

E) Catalytic Hydrogenation and Reduction of Alkynes

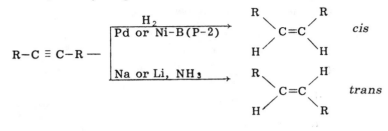

3.4 REACTIONS OF ALKENES

A) Hydroxylation (Glycol Formation)

$$\underset{/}{\overset{\backslash}{C}} = \underset{\backslash}{\overset{/}{C}} \xrightarrow[\text{or } HCO_2OH]{\text{cold alkaline } KMnO_4} \underset{\underset{OH}{|}}{-C} \underset{\underset{OH}{|}}{-C} -$$

B) Polymerization

Ex. $n\ CH_2 = CH_2 \xrightarrow[\text{pressure}]{O_2,\ \text{Heat}} (-CH_2-CH_2-)_n$

       ethylene   Polyethylene

C) Addition of Hydrogen Halides

$$-C=C- + HX \rightarrow -\underset{\underset{H}{|}}{C} \underset{\underset{X}{|}}{-C}- \qquad X = Cl,\ Br,\ I$$

In Markovnikov addition the hydrogen of the acid attaches itself to the carbon atom that has the greater number of hydrogens. In anti-Markovnikov addition the hydrogen of the acid attaches itself to the carbon atom that has the least number of hydrogens. Addition of hydrogen chloride or iodide follows Markovnikov's rule. Addition of hydrogen bromide will follow either rule, solely determined by the presence or absence of peroxides.

Ex. $CH_3CH = CH_2 \xrightarrow{HI} CH_3CHICH_3$

     propene   2-iodopropane

$CH_3CH = CH_2 \xrightarrow{HBr}$

- $\xrightarrow[\text{peroxides}]{\text{no}}$ $CH_3CHBrCH_3$ Markovnikov addition
  2-bromopropane
- $\xrightarrow{\text{peroxides}}$ $CH_3CH_2CH_2Br$ Anti-Markovnikov addition
  1-bromopropane

D) Hydroboration-Oxidation

$$-\overset{|}{C}=\overset{|}{C}- + (BH_3)_2 \rightarrow -\overset{|}{\underset{H}{C}}-\overset{|}{\underset{B}{C}}- \xrightarrow[OH^-]{H_2O_2} -\overset{|}{\underset{H}{C}}-\overset{|}{\underset{OH}{C}}-$$

              diborane

Anti-Markovnikov addition

E) Oxymercuration-Demercuration

$$-\overset{|}{C}=\overset{|}{C}- + H_2O + Hg(OAc)_2 \rightarrow -\overset{|}{\underset{OH}{C}}-\overset{|}{\underset{HgOAc}{C}}- \xrightarrow{NaBH_4} -\overset{|}{\underset{OH}{C}}-\overset{|}{\underset{H}{C}}-$$

Markovnikov addition

F) Addition of Halogens

$$-\overset{|}{C}=\overset{|}{C}- + X_2 \rightarrow -\overset{|}{\underset{X}{C}}-\overset{|}{\underset{X}{C}}- \qquad X = Cl_2, Br_2$$

Ex. $CH_3CH = CH_2 \xrightarrow{Br_2+CCl_4} CH_3CHBrCH_2Br$

     propene   1,2-dibromopropane

G) Catalytic Hydrogenation (Addition of Hydrogen)

$$-\overset{|}{C}=\overset{|}{C}- + H_2 \xrightarrow[\text{or Ni}]{Pt_1, Pd_1} -\overset{|}{\underset{H}{C}}-\overset{|}{\underset{H}{C}}-$$

22

Ex.

$$CH_3CH = CH_2 \xrightarrow{H_2,Ni} CH_3CH_2CH_3$$

propene   propane

3.5 DIENES

Dienes have the structural formula, C_nH_{2n-2}. In the IUPAC nomenclature system, dienes are named in the same manner as alkenes, except that the suffix "-ene" is replaced by "-diene," and two numbers must be used to indicate the position of the double bonds.

3.6 CLASSIFICATION OF DIENES

$$-\overset{|}{C} = C = \overset{|}{C}-$$ Cumulated double bonds (allenes)

$$-\overset{|}{C} = \overset{|}{C}-\overset{|}{C} = \overset{|}{C}-$$ Conjugated (alternating) double bonds

$$-\overset{|}{C} = \overset{|}{C}-(CH_2)\underset{n}{-}\overset{|}{C} = \overset{|}{C}-$$ Isolated (non-conjugated) double bonds

3.7 PREPARATION OF DIENES

All preparation methods used for the alkenes may be used for non-conjugated dienes using di-functional starting materials. Conjugated dienes may be produced in the following ways:

A) Dehydration of 1,3-Diols

$$CH_3-\underset{\underset{OH}{|}}{CH}-CH_2-\underset{\underset{OH}{|}}{CH_2} \xrightarrow[\text{Acid}]{\text{Heat}} CH_2 = CH-CH = CH_2 + 2H_2O$$

1,3-butanediol   1,3-butadiene

B) Dehydrogenation

$$CH_3-CH_2-CH_2-CH_3 \xrightarrow[\text{catalyst}]{\text{Heat}} \begin{cases} \rightarrow CH_3-CH_2-CH = CH_2 \\ CH_2 = CH-CH = CH_2 \\ \rightarrow CH_3-CH = CH-CH_3 \end{cases} \xleftarrow[\text{Catalyst}]{\text{Heat}}$$

Allene may be produced from glycerol by stepwise substitution and eliminations.

$$\begin{matrix} CH_2-OH \\ | \\ CH-OH \\ | \\ CH_2-OH \end{matrix} \xrightarrow{HBr} \begin{matrix} CH_2-Br \\ | \\ CH-Br \\ | \\ CH_2-Br \end{matrix} \xrightarrow[\text{Alc.}]{KOH} \begin{matrix} CH_2 \\ \| \\ C-Br \\ | \\ CH_2-Br \end{matrix} \xrightarrow[\text{alc.}]{Zn} \begin{matrix} CH_2 \\ \| \\ C \\ \| \\ CH_2 \end{matrix}$$

3.8 REACTIONS OF DIENES

A)   1,4-Additions to give 1,4-Alkadienes

Ex. $CH_2 = CH-CH = CH_2 + Br_2 \rightarrow$
$$\begin{matrix} CH_2-CH = CH-CH_2 \\ | \quad\quad\quad\quad | \\ Br \quad\quad\quad\quad Br \end{matrix}$$

   1,3-butadiene   1,4-dibromobutene

B)   Isomerization of certain Dienes to give Alkynes

Ex. $(CH_3)_2C = C = CH_2 + Na \rightarrow (CH_3)_2CH-C \equiv CH$

C)   Polymerization to give synthetic rubber

Ex.
$$n\,CH_2 = \overset{\overset{\displaystyle CH_3}{|}}{C}-CH = CH_2 \xrightarrow{\text{Catalyst}} \text{Rubber-like products}$$
   isoprene

$$n\,CH_2 = \overset{\overset{\displaystyle Cl}{|}}{C}-CH = CH_2 \longrightarrow \text{Neoprene}$$
   chloroprene

CHAPTER 4

ALKYNES

Alkynes are unsaturated hydrocarbons containing triple bonds. They have the general formula, C_nH_{2n-2}.

$$R-C \equiv C-R \qquad R = \text{Alkyl or H}$$

Ex. $H-C \equiv C-H$ Simplest alkyne

acetylene

4.1 NOMENCLATURE (IUPAC SYSTEM)

Alkynes are named in the same manner as alkenes, except that the suffix "-ene" is replaced with "-yne."

When both a double bond and a triple bond are present, the hydrocarbon is called an alkenyne. In this case, the double bond is given preference over the triple bond when numbering.

Ex. $CH_3-C \equiv CH$ $CH_3-CH \equiv CH-C \equiv CH$

propyne   1-penten-3-yne

4.2 PHYSICAL PROPERTIES OF ALKYNES

A) Lower-carbon members are gases with boiling points somewhat higher than corresponding alkenes.

B) Terminal alkynes have lower boiling points than isomeric internal alkynes.

C) The hydrogens in terminal alkynes are relatively acidic.

D) The dipole moment is small, but larger than that of an alkene.

E) Other physical properties are essentially the same as those for alkanes and alkenes.

4.3 PREPARATION OF ALKYNES

A) Dehydrohalogenation of Alkyl Dihalides

$$
\begin{array}{ccc}
\overset{H}{\underset{X}{-C}}\overset{H}{\underset{X}{-C-}} & \xrightarrow[\text{Alc.}]{\text{KOH}} & \overset{H}{-C=C-} \\
\end{array}
\underset{X}{} \xrightarrow{\text{NaNH}_2} -C \equiv C-
$$

Ex.

$$
CH_3-\underset{Br}{\underset{|}{CH}}-\underset{Br}{\underset{|}{CH_2}} \xrightarrow[\text{Alc.}]{\text{KOH}} CH_3-CH=\underset{Br}{\underset{|}{CH}} \xrightarrow{\text{NaNH}_2} CH_3C \equiv CH
$$

propyne

1,2-dibromopropane

B) Dehalogenation of Tetrahalides

$$
\underset{\underset{X}{\underset{|}{X}}}{\overset{\overset{X}{\overset{|}{X}}}{-C-C-}} + 2Zn \rightarrow -C \equiv C- + 2ZnX_2
$$

Ex.

$$
CH_3-\underset{\underset{Br}{\underset{|}{Br}}}{\overset{\overset{Br}{\overset{|}{Br}}}{C-CH}} \xrightarrow{2Zn} CH_3-C \equiv CH + 2ZnBr_2
$$

1,1,2,2-tetrabromopropane   propyne

C) Reaction of Water with Calcium Carbide

$$CaC_2 + H_2O \rightarrow CH \equiv CH + Ca(OH)_2$$

$$\text{acetylene}$$

4.4 REACTIONS OF ALKYNES

A) Addition of Hydrogen

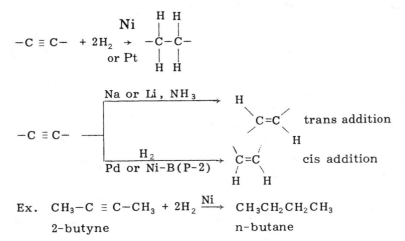

Ex. $CH_3-C \equiv C-CH_3 + 2H_2 \xrightarrow{Ni} CH_3CH_2CH_2CH_3$

2-butyne n-butane

B) Addition of Halogens

$-C\equiv C- \xrightarrow{X_2} -C=C- \xrightarrow{X_2} -\overset{X}{\underset{X}{C}}-\overset{X}{\underset{X}{C}}-$ $X_2 = Cl_2, Br_2$

Ex.

$CH_3-C \equiv CH \xrightarrow{Br_2} CH_3-C=CH \xrightarrow{Cl_2} CH_3-\overset{Cl}{\underset{Br}{C}}-\overset{Cl}{\underset{Br}{C}}H$

C) Addition of Hydrogen Halides

$$-C \equiv C- \xrightarrow{HX} \underset{\overset{|}{H}\quad\overset{|}{X}}{-C=C-} \xrightarrow{HX} \underset{\overset{|}{H}\quad\overset{|}{X}}{\overset{\overset{|}{H}\quad\overset{|}{X}}{-C-C-}} \qquad X = Cl, Br, I$$

Ex.

$$CH_3 C \equiv CH \xrightarrow{HCl} \underset{\overset{|}{Cl}}{CH_3 C = CH_2} \xrightarrow{HI} \underset{\overset{|}{Cl}}{\overset{\overset{|}{I}}{CH_3 - C - CH_3}}$$

D) Addition of Water (Hydration)

$$-C \equiv C- \; + H_2O \xrightarrow{H_2SO_4, HgSO_4} \underset{\overset{|}{H}\quad\overset{|}{OH}}{-C = C--} \; \overset{\leftarrow}{\longrightarrow} \; \underset{\overset{|}{H}\quad\overset{\|}{O}}{-C-C-}$$

Ex.

$$CH_3 - C \equiv CH + H_2O \xrightarrow[HgSO_4]{H_2SO_4} \underset{\overset{|}{H}\quad\overset{\|}{O}\quad\overset{|}{H}}{\overset{\overset{|}{H}\qquad\overset{|}{H}}{H - C - C - C - H}}$$

propyne   2-propanone (acetone)

E) Nucleophilic Additions

Unlike the simple alkenes, alkynes undergo these additions.

Ex. Reaction with alkoxides in alcoholic solution to yield vinyl ethers.

$$CH \equiv CH + RO^- \xrightarrow[\substack{\text{High Temp.} \\ \text{and Press.}}]{ROH} ROCH = CH^- \xrightarrow{ROH}$$

$$ROCH = CH_2 + RO^-$$

CHAPTER 5

ALKYL HALIDES

Alkyl halides are compounds in which one hydrogen atom is replaced by an atom of the halide family. An important use of alkyl halides is as intermediates in organic synthesis.

Structural formula: $C_nH_{2n+1}-X$; $X = Cl, Br, I, F$.

5.1 NOMENCLATURE (IUPAC SYSTEM)

Table 5.1

Formula	Name
CH_3Cl	chloromethane
CH_3CH_2Br	bromoethane
$CH_3CH_2CH_2I$	1-iodopropane
CH_3CHICH_3	2-iodopropane
$CH_3CH_2CH_2CH_2Cl$	1-chlorobutane
$CH_3CH_2CHBrCH_3$	2-bromobutane
$(CH_3)_3CI$	2-iodo-2-methylpropane
$CH_3CH_2CH_2CH_2CH_2Cl$	1-chloropentane

5.2 PHYSICAL PROPERTIES OF ALKYL HALIDES

The specific gravities and boiling points increase with increasing atomic weight of the halogen atom.

Most monohaloalkanes up to C_{18} are liquids at room temperature.

Alkyl halides are usually polar molecules.

Boiling points of alkyl halides are much higher than those of alkanes with the same carbon skeleton.

Alkyl halides are insoluble in water, but are soluble in the typical organic solvents.

They are good solvents for most organic compounds.

They are colorless when pure and have a pleasant odor.

5.3 PREPARATION OF ALKYL HALIDES

A) From Alcohols

  a) Hydrogen chloride or bromide in the presence of sulfuric acid or zinc chloride.

$$R-OH + HX \xrightarrow{H_2SO_4} R-X + H_2SO_4 \cdot H_2O$$

  R may rearrange

  Ex. $CH_3CH_2CH_2OH + HBr \xrightarrow{H_2SO_4} CH_3CH_2CH_2Br + H_2SO_4$
  $$H_2O$$

  1-propanol   1-bromopropane

  b) Dry hydrogen chloride and bromide

  $$R-OH + HX, \text{ dry} \rightarrow R-X + H_2O$$

  Ex. $CH_3CH_2OH + HBr, \text{ dry} \rightarrow CH_3CH_2Br + H_2O$

  ethanol   bromoethane

  c) Phosphorous Halides

  $$R-OH + PX_5 \rightarrow R-X + POX_3 + HX$$

  and

  $$3R-OH + PX_3 \rightarrow 3R-X + H_3PO_3$$

Ex. $CH_3CH_2OH + PCl_5 \rightarrow CH_3CH_2Cl + POCl_3 + HCl$

ethanol$\qquad$chloroethane

d) Thionyl Chloride

$$R-OH + SOCl_2 \rightarrow R-Cl + SO_2 + HCl$$

B) Addition of hydrogen halides to unsaturated hydrocarbons

a) $H_2C = CH_2 + HX \rightarrow CH_3-CH_2-X$ primary halide

b) $R-HC = CH_2 + HX \rightarrow R-CHX-CH_3$ secondary halide

c) $R_2C = CH_2 + HX \rightarrow R_2CX-CH_3$ tertiary halide

C) Halogenation of alkanes

$$R-H + X_2 \rightarrow R-X + HX$$

Ex. $CH_4 + Cl_2 \rightarrow CH_3Cl + HCl$

$\qquad\qquad$ chloromethane

D) Halide Exchange

a) $R-X + NaI \xrightarrow{\text{acetone}} R-I + NaX$ $X = Cl, Br$

b) $R-X + AgF \longrightarrow R-F + AgX$

c) $R-CCl_3 + SbF_3 \longrightarrow R-C-F_3 + SbCl_3$

HALOFORM REACTION

Methyl ketones are converted to acids,

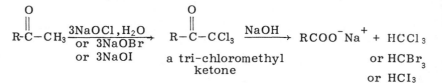

HALOGENATION OF ALKENES

Electrophilic addition of X_2 ($X = Br, Cl$) yields 1,2-di-halides.

$$-\overset{|}{\underset{|}{C}} = \overset{|}{\underset{|}{C}} - + X_2 \rightarrow -\overset{|}{\underset{\underset{X}{|}}{C}} - \overset{|}{\underset{\underset{X}{|}}{C}} -$$

HUNSDIECKER REACTION

Long-chain alkyl bromides are obtained from fatty acids.

$$R-(CH_2)_n COO^- Ag^+ \xrightarrow[\text{heat}]{Br_2, CCl_4} R-(CH_2)_n Br + CO_2 + AgBr$$

5.4 REACTIONS OF ALKYL HALIDES

Alkyl halides undergo nucleophilic substitution as shown in Table 5-2.

Reactions of alkyl halides with Grignard reagents produce alkanes.

$$RX + Mg \xrightarrow{\text{dry ether}} RMgX$$

$$R'X + RMgX \longrightarrow R-R' + MgX_2$$

Alkyl halides undergo double decomposition reactions when treated with aqueous sodium (or potassium) hydroxide, sodium (or potassium) alkoxides, sodium (or potassium) salts of fatty acids, sodium (potassium, silver) cyanide, sodium (potassium; silver) nitrite, and silver oxide.

$$R-X + NaOH, aq. \rightarrow R-OH + NaX$$

$$R-X + NaOOCR \rightarrow R-OOCR + NaX$$

$$R-X + NaOR \rightarrow R-OR + NaX$$

$$R-X + NaCN \rightarrow R-CN + NaX$$

$$R-X + AgCN \rightarrow R-CN + AgX$$

$$R-X + NaONO \rightarrow R-ONO + NaX$$

$$R-X + AgO \rightarrow R-OH + AgX$$

Table 5-2   Nucleophilic Substitutions of Alkyl Halides

Nucleophile		Product	
R−X + :ÖH⊖	Hydroxide	R−OH	Alcohol
:ÖH$_2$	Water	R−OH	Alcohol
:ÖR⊖	Alkoxide	R−OR	Ether (Williamson)
⊖OOC−R'	Carboxylate	R−OOC−R'	Ester
:ṢH⊖	Hydrosulfide	R−SH	Thiol
:ṢR⁻	Thioalkoxide	R−SR'	Sulfide
:ṢR'$_2$	Sulfide	R−ṢR'$_2$⊕ X⊖	Sulfonium salt
SCN⊖	Thiocyanide	R−SCN	Alkyl thiocyanide
:Ï:⊖	Iodide	R−I	Alkyl iodide
:ṄH$_2$⊖	Amide	R−NH$_2$	1° Amine
:NH$_3$	Ammonia	R−NH$_2$	1° Amine
:NH$_2$R'	1° Amine	R−NHR'	2° Amine
:NHR'$_2$	2° Amine	R−NR'$_2$	3° Amine
:NR'$_3$	3° Amine	R−NR'$_3$⊕ X⁻	Quaternary ammonium salt
N$_3$⊖	Azide	R−N$_3$	Alkyl azide
NO$_2$⊖	Nitrite	R−NO$_2$	Nitroalkane
:P(C$_6$H$_5$)$_3$	Phosphine	R−P(C$_6$H$_5$)$_3$⊕ X⁻	Phosphonium salt
⊖:C≡N:	Cyanide	R−CN	Nitrile
⊖:C≡C−R'	Alkynyl anion	R−C≡C−R'	Alkyne
⊖:R'	Carbanion	R−R'	Alkane
⊖:CH(COOR')$_2$		R−CH (COOR')$_2$	Malonic ester synthesis
⊖:CH(COCH$_2$)(COOR)		R—CH (COCH$_2$)(COOR)	Acetoacetic ester synthesis
Ar−H , AlCl$_3$		R−Ar	Alkyl benzene (Friedel-Crafts)

5.5 NUCLEOPHILIC DISPLACEMENT REACTIONS

Alkyl halides undergo nucleophilic substitution (S_N) and elimination (E) in the presence of basic reagents. The two reactions are always in competition.

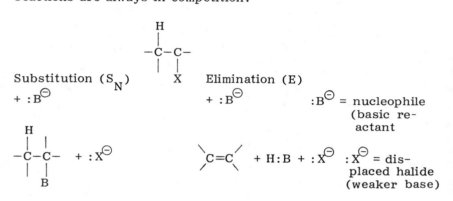

In the S_N reaction the base, $:B^\ominus$, replaces the weaker base, $:X^\ominus$. The E reaction is the reverse of an addition reaction and a hydrogen and a halogen on adjacent carbons are eliminated. Typical bases were shown in Table 5-2.

5.6 $S_N 1$ AND $S_N 2$ SUBSTITUTION REACTIONS

A nucleophilic substitution reaction (S_N reaction) is the typical reaction encountered by an alkyl halide in the presence of a basic electron-rich reagent (nucleophile).

$$R:X + :B^\ominus \rightarrow R:B + :X^\ominus$$

Ex. $$CH_3Br + :OH^\ominus \rightarrow CH_3OH + :Br^\ominus$$

The mechanism of the above bimolecular S_N reaction

(S_N2) involves a direct collision between the nucleophile (:OH$^-$) and the carbon bearing the halide. The S_N2 reaction follows second order kinetics because its rate depends upon the concentrations of the two reacting substances.

Inversion of configuration has taken place when a reaction yields a product whose configuration is opposite to that of the reactant. The S_N2 reaction causes complete stereochemical inversion due to backside attack.

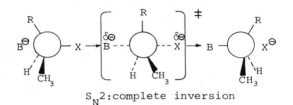

S_N2: complete inversion

The reactivity toward S_N2 substitution decreases as the number of substituents attached to the carbon carrying the halogen increases. The order of reactivity is

$$CH_3X > 1° > 2° > 3°$$

Another example of a nucleophilic substitution reaction is:

$$CH_3-\underset{\underset{Br}{|}}{\overset{\overset{CH_3}{|}}{C}}-CH_3 + :OH^{\ominus} \rightarrow CH_3-\underset{\underset{OH}{|}}{\overset{\overset{CH_3}{|}}{C}}-CH_3 + :Br^{\ominus}$$

The mechanism of the above unimolecular S_N reaction (S_N1) does not involve collision because the rate determining step involves only one molecule. The rate determining step is the single step whose rate determines the overall rate of a stepwise reaction. The rate of the entire reaction is determined by how fast the alkyl halide ionizes, and it therefore depends upon only the concentration of the alkyl halide. As a result the S_N1 reaction follows first-order kinetics.

The S_N1 reaction is characterized by the formation of a carbonium ion. The carbonium ion may be attacked by the hydroxide ion from either side of its plane so as to cause either an inversion or retention of configuration.

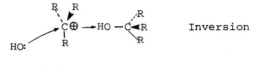

Inversion

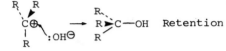

Retention

S_N1: Inversion and Retention

In contrast to an S_N2 reaction, the S_N1 reaction proceeds with racemization.

In S_N1 reactions the order of reactivity is allyl, benzyl > 3° > 2° > 1° > CH_3X.

The rate of S_N2 reactions is affected by steric factors (the bulk of the substituents). The rate of S_N1 reactions is affected by electronic factors (the tendency of substituents to release or withdraw electrons). Electron-donating groups favor the S_N1 mechanism, in which a positive charge is generated. Electron-withdrawing groups favor the S_N2 mechanism, in which the transition state is more negative than the starting material.

$\xrightarrow{\quad}$	$\xleftarrow{\quad}$	$\xleftarrow{\quad}$	$\xrightarrow{\quad}$
$\delta +$	$\delta +$	$\delta -$	$\delta -$
Z———C	Z———C	Z———C	Z———C

Favorable	Unfavorable	Favorable	Unfavorable
Charge interaction in		Charge interaction in	
S_N1 transition state		S_N2 transition state	

High concentrations of the nucleophilic reagent favor S_N2 reaction; low concentrations favor S_N1 reaction.

The polarity of the solvent also determines the mechanism by which the reaction occurs. Increasing solvent polarity favors the S_N1 reaction (slows down the S_N2 reaction).

36

Table 5-3

Compound	Structure	S_N Mechanism
Primary Alkyl Halides	RCH_2-X	S_N2
Secondary Alkyl Halides	R_2CH-X	Either S_N1 or S_N2, depending upon the solvent present.
Tertiary Alkyl Halides	R_3C-X	S_N1 occurs when an ionizing solvent is present. A very slow S_N2 occurs when a non-ionizing solvent is used.

The effects of the nucleophile on reaction rate are as follows:

A) The nucleophilicity for a given atom increases with an increase in the negative charges: $:HO^{\ominus}$ is more nucleophilic than H_2O.

B) An increase in the atomic number within a row of the periodic table decreases nucleophilicity: $C^- > N^- > O^- > F^-$, or $NH_3 > H_2O > HF$.

C) An increase in the atomic number within a column of the periodic table increases nucleophilicity: $I^- > Br^- > Cl^- > F^-$.

5.7 ELIMINATION REACTION: E2 AND E1

E2

Bimolecular Elimination

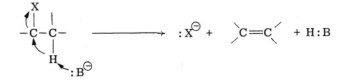

37

E1

Unimolecular Elimination

Rate determining step

(1)

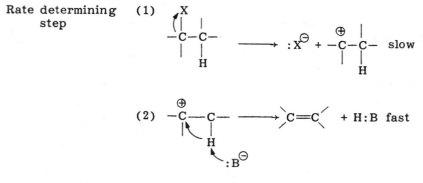

The E2 mechanism (elimination, bimolecular) involves two molecules in the rate-determining step.

The E1 mechanism (elimination, unimolecular) involves one molecule in the rate-determining step.

Reactivity toward   3° > 2° > 1°
E2 or E1 Elimination

The E1 elimination mechanism follows first-order kinetics, demonstrates the identical effect of the structure reactivity, and can be accompanied by rearrangement of configuration.

The E2 elimination mechanism follows second-order kinetics, is not accompanied by rearrangements, demonstrates a large deuterium isotope effect, does not undergo hydrogen-deuterium exchange, and demonstrates a large element effects.

There is a variable E2 elimination mechanism involving the formation of a carbanion.

(1)

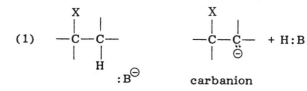

carbanion

(2)

$$\underset{\underset{\ominus}{\overset{X}{\underset{|}{\overset{|}{-C-C-}}}}}{} \longrightarrow :X^- + \overset{\diagdown}{\underset{\diagup}{C}}=\overset{\diagup}{\underset{\diagdown}{C}}$$

Variable E2 Mechanism

The mechanism that will be operative in a given situation depends primarily on the structure of the substrate, the nature of the solvent, the strength of the base, and the nature of the leaving group.

The ionization of a carbon-hydrogen bond occurs more readily if the carbon atom is highly branched. Ionization proceeds only if an ionizing solvent is present.

The order of decreasing reactivity in ionization is

tertiary > Secondary > Primary > methyl,vinyl

5.8 STEREOCHEMISTRY OF ELIMINATION

The E2 elimination yields two pairs of enantiomers, cis and trans. This elimination mechanism is stereospecific.

Example

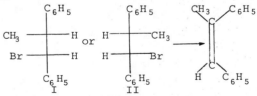

1-Bromo-1,2-diphenylpropane   cis-1,2-diphenyl-1-propene

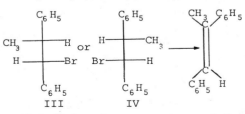

1-Bromo-1,2-diphenylpropane   trans-1,2-diphenyl-1-propene

39

Anti-elimination is involved in the bimolecular reaction of alkyl halides. The hydrogen and the leaving group, in its transition state, are far apart from each other.

The anti-relationship is shown as follows:

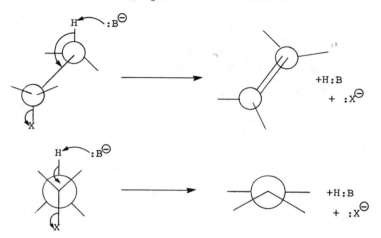

Syn-elimination is also involved in the E2 reactions. The hydrogen and the leaving group, in the transition state, are in the eclipsed (or gauche) relationship.

The anti-elimination mechanism is the more stable mechanism, because the anti-conformation is far more stable than the eclipsed conformation.

CHAPTER 6

STEREOCHEMISTRY-
STEREOISOMERISM

6.1 OPTICAL ACTIVITY

Plane-polarized light is light whose vibrations take place in only one of the infinite planes through its line of propagation.

An optically active substance is one that rotates the plane of polarized light.

The rotation is measured and detected by an instrument called a polarimeter. The amount of rotation is dependent upon the number of molecules the light meets while passing through the polarimeter tube. Substances which rotate the plane of polarized light to the right are called dextrorotatory and are symbolized by d or +. Their mirror images which rotate light to the left are called levorotatory and are symbolized by l or -.

Specific rotation is the number of degrees of rotation observed if a 1-decimeter tube is used, and the compound being examined is present to the extent of 1g/cc.

$$[\alpha] = \frac{\alpha}{1 \times d}$$

$$\text{specific rotation} = \frac{\text{observed rotation (degrees)}}{\text{length (dm)} \times \text{g/cc}}$$

where d represents density for a pure liquid or concentration for a solution.

6.2 CHIRALITY

A chiral center (C*) is a carbon atom with four different groups attached to it.

The following are some examples of chiral carbons:

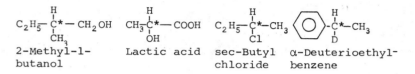

2-Methyl-1-butanol   Lactic acid   sec-Butyl chloride   α-Deuterioethyl-benzene

Chiral molecules are not superimposable on their mirror images. Not all molecules that contain a chiral center are chiral, and not all chiral molecules contain a chiral center. Chiral molecules do not have a plane of symmetry.

The maximum number of stereisomers that are possible for a compound with n chiral centers is given by 2^n.

6.3 ENANTIOMERISM

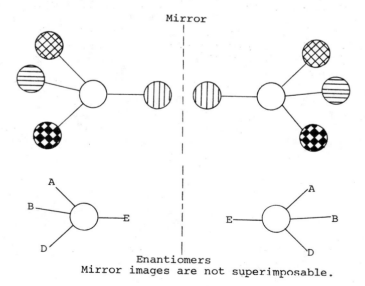

Enantiomers
Mirror images are not superimposable.

Enantiomers are isomers that are mirror images of each other, and they belong to the general class called stereoisomers. Stereoisomers are isomers that are different from each other only in the way the atoms are oriented in space.

Chiral molecules can exist as enantiomers, and achiral (without chirality) molecules cannot exist as enantiomers.

Enantiomers have identical physical properties except for the direction in which they rotate plane-polarized light. Enantiomers have identical chemical properties except toward optically active reagents.

6.4 DIASTEREOMERS

Stereoisomers that are not mirror images of each other are called diastereomers.

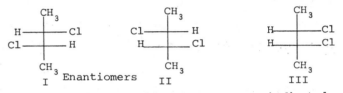

III is a diastereomer of I and II where indicated

Diastereomers have similar (not identical) chemical properties, since they are members of the same family.

Diastereomers have different physical properties and differ in specific rotation; they may be levorotatory, dextrorotatory, or inactive.

The particular class of diastereomers that results from restricted rotation about a double bond are called geometric isomers.

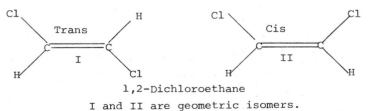

1,2-Dichloroethane

I and II are geometric isomers.

43

Geometric isomers have different physical properties and can be separated from each other.

A meso compound is one whose molecules are superimposable on their mirror images even though they contain chiral centers. A meso compound is optically inactive.

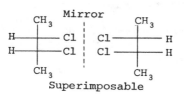

We can often recognize a meso compound on sight by the fact that half the molecule is the mirror image of the other half.

6.5 RACEMIZATION

Racemization is the process that leads to the formation of a racemic modification, a mixture of equal parts of enantiomers.

A racemic modification is optically inactive because the rotating influence of one enantiomer just cancels that of the other.

The separation of a racemic modification into its component enantiomers is called resolution.

There are three methods commonly used to carry out resolutions: mechanical separation of asymmetric crystals, resolution by formation of diastereomers, and resolution by reaction with optically active reagents. The last method mentioned is the most widely used.

6.6 SPECIFICATION OF CONFIGURATION

The arrangement of atoms that characterizes a particular stereoisomer is called its configuration.

For specifying particular configurations we follow two steps:

Step 1. A sequence of priority according to rules is assigned to the four atoms or groups of atoms attached to the chiral carbon.

Step 2. The molecule is oriented so that the group of lowest priority is directed away from us. The arrangement of the remaining groups is then observed. If our eye travels in a clockwise direction in going from the group of highest priority to the groups of lower priority, the configuration is specified by R. If eye movement is counterclockwise, the configuration is specified by S.

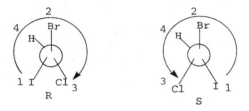

Sequence Rule 1. If the four atoms attached to the chiral center are all different, then priority depends on atomic number, with the atom of higher atomic number having priority. In the case of two atoms being isotopes of the same element, the atom of higher mass number has priority.

Sequence Rule 2. If priority cannot be decided from the first rule, then it will be determined by similar comparison of the atoms attached to it, and so on.

Sequence Rule 3. Both atoms are considered to be duplicated or triplicated when there exists between them a double or triple bond.

$$-\overset{|}{C}=O \text{ equals } -\overset{|}{\underset{|}{C}}-\overset{}{\underset{|}{O}} \quad \text{and} \quad -C \equiv N \text{ equals } -\overset{N\quad C}{\underset{N\quad C}{\overset{|\quad|}{C}-N}}$$
$$O \quad C$$

Configurations for compounds with more than one chiral center are specified by specifying the configuration for each of the centers and by using the number of the chiral center.

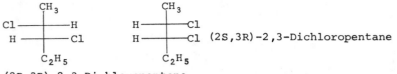

(2R,3R)-2,3-Dichloropentane   (2S,3R)-2,3-Dichloropentane

FISCHER CONVENTION

D or L isomers are used to represent the Fischer formula. D or L isomers are either positive or negative depending on the particular compound involved.

An isomer belongs to the D series if, when it is written with the aldehyde-containing group (or a related group such as COOH) attached to the chiral carbon at the top of the molecule and with the other carbon-containing group attached to the chiral carbon at the bottom, the OH group is on the right side. Its enantiomer belongs to the L series. When a compound has more than one chiral carbon atom, the compound belongs to either the D or the L series based on the configuration at the highest-numbered chiral carbon atom (farthest removed from the COH group).

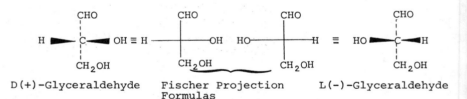

D(+)-Glyceraldehyde   Fischer Projection Formulas   L(-)-Glyceraldehyde

CHAPTER 7

CYCLIC HYDROCARBONS

Cyclic alkanes and cyclic alkenes are alicyclic (aliphatic cyclic) hydrocarbons.

7.1 NOMENCLATURE

Cyclic aliphatic hydrocarbons are named by prefixing the term "cyclo-" to the name of the corresponding open-chain hydrocarbon, having the same number of carbon atoms as the ring.

Ex.

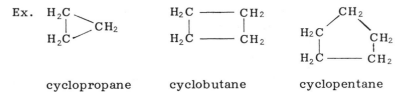

cyclopropane   cyclobutane   cyclopentane

Substituents on the ring are named, and their positions are indicated by numbers, the lowest combination of numbers being used.

7.2 PROPERTIES OF CYCLIC HYDROCARBONS

Cyclopropane and cyclobutane are both colorless gases but cyclopentane is a colorless liquid.

The boiling points of the cycloalkanes are about $10°$-$20°$ higher than those of corresponding open-chain alkanes.

Their densities increase with an increase in the carbon chain length.

Cycloalkanes and cycloalkenes are insoluble in water but soluble in alcohol and ether.

The melting points of cyclic hydrocarbons are higher than those of alkanes because the cyclic hydrocarbons fit more readily into a crystal lattice.

The heat of combustion per methylene group depends on the ring size.

7.3 PREPARATION OF CYCLIC HYDROCARBONS

Reaction of active metals (Na, Mg, Zn, etc.) with certain dihalogenated hydrocarbons.

a) $X - (CH_2)_n - X + 2M(\text{or } M) \rightarrow (CH_2)_n + 2MBr \text{ (or } MBr_2)$

cycloalkane

Ex. $BrCH_2CH_2CH_2Br + Zn \rightarrow$

cyclopropane

b) Partial or complete reduction of benzene in the presence of heated nickel results in the formation of cyclohexadiene, cyclohexene, or cyclohexane.

Ex. $C_6H_6 + 3H_2 \xrightarrow[\text{Heat} < 300°C]{\text{finely divided Ni}} C_6H_8 \rightarrow C_6H_{10} \rightarrow C_6H_{12}$

Hydrogenation of Arenes

Addition of carbenes to alkenes

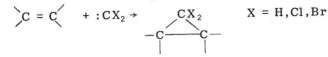

$\diagup C = C \diagdown + :CX_2 \rightarrow$ $\underset{-C—C-}{CX_2}$ $X = H, Cl, Br$

7.4 REACTIONS OF CYCLOALKANES AND CYCLOALKENES

Reaction of bromine with cyclopentane to give bromocyclopentane.

Ex. $\underset{CH_2——CH_2}{\overset{CH_2——CH_2}{|}} CH_2 + Br_2 \xrightarrow{300°C} \underset{CH_2CH_2}{\overset{CH_2CH_2}{|}} CHBr + HBr$

bromocyclopentane

Cycloalkanes undergo free radical substitution

Ex. $\underset{H_2C}{\overset{H_2C}{|}} CH_2 + Cl_2 \xrightarrow[heat]{light} \underset{H_2C}{\overset{H_2C}{|}} CHCl + HCl$

cyclopropane   chlorocyclopropane

Cycloalkenes undergo both electrophilic and free-radical addition reactions; they also undergo cleavage and allylic substitutions.

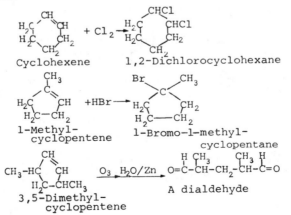

Cyclohexene   1,2-Dichlorocyclohexane

1-Methyl-cyclopentene   1-Bromo-1-methyl-cyclopentane

3,5-Dimethyl-cyclopentene   A dialdehyde

Chain addition reactions of cyclopropane and cyclobutane: These addition reactions destroy the cyclopropane and cyclobutane ring systems and yield open chain products:

$$
\underset{H_2C}{\overset{H_2C}{\diagdown}}\!\!\diagup CH_2
$$

Ni, H$_2$ 80°C → CH$_2$-CH$_2$-CH$_2$ (with H, H) (Propane)

Cl$_2$, FeCl$_3$ → CH$_2$-CH$_2$-CH$_2$ (with Cl, Cl) (1,3-Dichloropropane)

Conc. H$_2$SO$_4$ → CH$_2$-CH$_2$-CH$_2$ (with H, OH) (1-Propanol)

7.5 BAEYER STRAIN

When carbon is bonded to four other atoms, the angle between any pair of bonds is a tetrahedral angle of 109.5°. But the ring of cyclopropane is a triangle with three angles of 60°. These deviations of bond angles from the "normal" value cause the molecule to be strained and hence unstable. The more unstable the molecule is more prone to undergo ring opening reactions.

7.6 CONFORMATION OF CYCLOALKANES

FACTORS AFFECTING STABILITY OF CONFORMATIONS

An angle strain (Baeyer strain) accompanies deviations from the "normal" bond angles (tetrahedral, 109.5° angle).

Carbon atoms in a tetrahedral arrangement tend to assume a staggered conformation.

A torsional strain (Pitzer strain) accompanies deviations from staggered arrangement.

The distance between non-bonding atoms is equal to the sum of their van der Waals radii. This causes them to attract each other.

When the attracted atoms are brought closer together, they repel each other, and a van der Waals strain (steric strain) results.

Non-bonded atoms (or groups of atoms) tend to take positions that result in the most favorable dipole-dipole interactions.

A molecule accepts a certain amount of angle strain to relieve van der Waals strain or dipole-dipole interactions.

CONFORMATIONS OF CYCLOALKANES

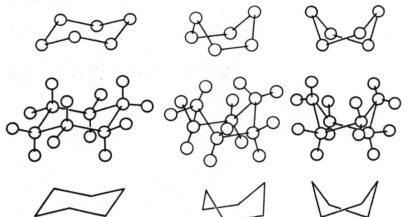

Chair Conformation   Boat Conformation   Twist-boat
                       (an energy maximum)   Conformation

Conformations of cyclohexane that are free of angle strain.

Along each of the carbon-carbon bonds in the chair form, there are perfectly staggered bonds.

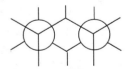

Chair cyclohexane   Staggered ethane

The chair form is the most stable conformation of cyclohexane and of nearly every derivative of cyclohexane. The chair form is a conformational isomer since it lies at an energy minimum.

Along each of the two carbon-carbon bonds in the boat conformation, there are sets of exactly eclipsed bonds.

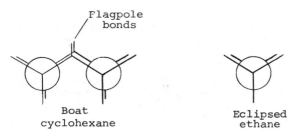

Boat
cyclohexane

Eclipsed
ethane

The chair conformation is much more stable than the boat conformation. The boat conformation is not a conformer, but a transition state between two conformers since it lies at an energy maximum.

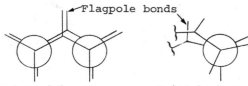

Boat cyclohexane

Twist-boat cyclohexane

The twist-boat conformation is a conformer since it lies at an energy minimum at 5.5 kcal above the chair conformation.

EQUATORIAL AND AXIAL BONDS IN CYCLOHEXANE

The equatorial bonds that hold the hydrogen atoms in the plane of the ring lie about the "equator" of the ring. The axial bonds that hold the hydrogen atoms above and below the plane are pointed along an axis perpendicular to the plane.

Equatorial Bonds

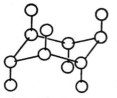

Axial Bonds

1,3 diaxial interaction results from the severe crowding among atoms that are held by the three axial bonds on the same side of the molecule. A given atom or group, except for hydrogen, has less room in an axial position than in an equatorial position.

7.7 STEREOISOMERISM OF CYCLIC COMPOUNDS

Stereoisomerism of cyclic compounds: cis- and trans-isomers.

cis-1,2-Cyclopentanediol

trans-1,2-Cyclopentanediol

The above isomers are configurational isomers; they are isolable because they are interconverted only by the breaking of bonds.

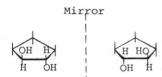

Not superimposable
Enantiomers:resolvable
trans-1,2-Cyclopentanediol

The trans-glycol is chiral, and its isomers are resolvable into optically active compounds (enantiomers).

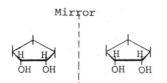

Superimposable
A meso compound
cis-1,2-Cyclopentanediol

The cis-glycol is a meso compound. **By definition its** mirror images are identical and unresolvable. The compound is not chiral and is optically inactive.

Stereoisomerism of Cyclic Compound Conformational Analysis

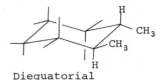
Diequatorial

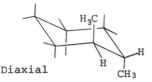
Diaxial

Chair conformations of trans-1,2-dimethylcyclohexane.

The diequatorial conformation is the more stable one because there is less crowding between $-CH_3$ groups and axial hydrogens of the ring (less 1,3 diaxial interaction).

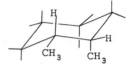
Equatorial-axial

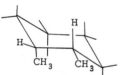
Axial-equatorial

Chair conformations of cis-1,2-dimethylcyclohexane.

The two cis- conformations are of equal stability and are less stable than the trans- isomer. Since in either cis-conformation, one $-CH_3$ group has to be in the axial position (1,3-diaxial interaction).

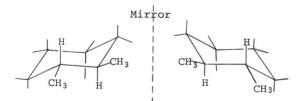

Not superimposable: not interconvertible
trans-1,2-Dimethylcyclohexane
A resolvable racemic modification

The trans-1,2-dimethylcyclohexane mirror images are not superimposable and therefore are enantiomers; they are not interconvertible and could be resolved into the enantiomers, each of which is optically active.

54

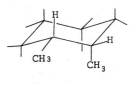 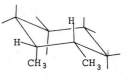

Not superimposable; but interconvertible
cis-1,2-Dimethylclohexane
A non-resolvable racemic modification

The cis-1,2-dim ethycyclohexane mirror images are not superimposable, and therefore, are enantiomers. In this case the enantiomers are interconvertible and cannot be resolved. These are conformational enantiomers.

CHAPTER 8

AROMATIC HYDROCARBONS

Most aromatic hydrocarbons (arenes) are derivatives of benzene. Examples of benzene derivatives are napthalene, anthracene and phenanthrene.

8.1 STRUCTURE

Benzene has a symmetrical structure and the analysis, synthesis and molecular weight determination indicate a molecular formula of C_6H_6.

Napthalene structure is indicated by the oxidation of 1-nitronapthalene which shows that the substituted ring is a true benzene ring. Reduction and oxidation of the same nucleus indicates that the unsubstituted ring is a true benzene ring.

8.2 NOMENCLATURE (IUPAC SYSTEM)

Aromatic compounds are named as derivatives of the corresponding hydrocarbon nucleus.

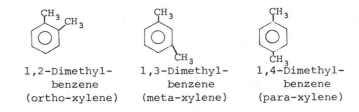

| 1,2-Dimethyl-
benzene
(ortho-xylene) | 1,3-Dimethyl-
benzene
(meta-xylene) | 1,4-Dimethyl-
benzene
(para-xylene) |

In the IUPAC system of nomenclature, the position of the substituent group is always indicated by numbers arranged in a certain order:

Benzene

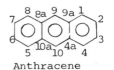

Naphthalene   Anthracene

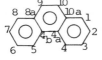

Phenanthrene

8.3 PREPARATION OF BENZENE AND ITS DERIVATIVES

PREPARATION OF BENZENE

Passage of acetylene and alkanes (CH_4 to C_6H_{14}) through hot tubes.

Ex. $3H-C \equiv C-H \xrightarrow[580°C]{\text{through tube}} C_6H_6$ (high yield)

$$CH_3-CH_2-CH_2-CH_2-CH_2-CH_3 \xrightarrow[\text{Cr}_2O_3]{\text{hot tower}} C_6H_6$$

Heating of phenol with zinc dust

$$C_6H_6OH + Zn, \text{ dust } \rightarrow C_6H_6 + ZnO$$

Hydrolysis of benzene sulfonic acid with superheated steam.

$$C_6H_5SO_2OH + H_2O, \text{ superheated} \atop \text{steam} \xrightarrow{\text{catalyst}} C_6H_6 + H_2SO_4$$

PREPARATION OF TOLUENE

The Wurtz Reaction - the action of sodium on a mixture of halobenzene and methylhalide.

Ex. $C_6H_5Br + 2Na + Br - CH_3$, ether $\rightarrow C_6H_5CH_3 + 2NaBr$

The Friedel-Crafts Reaction - action of an alkylhalide on benzene in the presence of anhydrous aluminum chloride.

Ex. $C_6H_5H + Br - CH_3 \xrightarrow{\text{anhydrous}}_{AlCl_3} C_6H_5 - CH_3 + HBr$

PREPARATION OF XYLENE

Reaction of a methyl halide with toluene in the presence of anhydrous aluminum chloride (Friedel-Crafts Reaction).

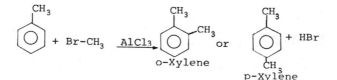

PREPARATION OF NAPHTHALENE

-Passage of benzene and acetylene through a hot tube.

Reaction of zinc or sodium with a mixture of 1,2-Bis bromomethylbenzene and 1,1,2,2 tetrabromoethane.

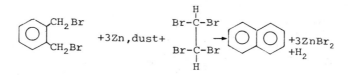

58

PREPARATION OF ANTHRACENE

Friedel-Crafts Reaction of benzene with 1,1,
2,2- tetrabromoethane in the presence of anhydrous aluminum chloride:

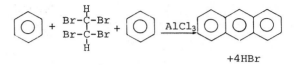

+4HBr

action of zinc dust with phenyl o-tolyl ketone:

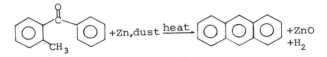

PREPARATION OF PHENANTHRENE

action of sodium with 1-bromo-2-(bromomethyl)
benzene.

8.4 PROPERTIES OF BENZENE

A) Colorless liquid

B) Boils at 80.1°C and melts at 5.5°C

C) **Symmetric benzene molecule has no net dipole moment.**

D) Completely soluble in all common organic solvents and insoluble in water.

E) In general, benzene does not react or behave like alkenes.

F) Benzene undergoes mainly substitution rather than addition reactions.

G) Benzene has the tendency to retain its conjugated unsaturated ring system (highly stable to chemical reagents).

The last three properties are due to benzene's aromaticity.

8.5 REACTIONS OF BENZENE, NAPHTHALENE, AND ANTHRACENE

SUBSTITUTION REACTIONS OF BENZENE

Nitration. Generation of electrophilic $^{\oplus}NO_2$ (nitronium ion) to attack nucleophilic benzene.

$$C_6H_6 + HONO_2 \xrightarrow{H_2SO_4} C_6H_5NO_2 + H_2O$$

Nitric acid   Nitrobenzene

Sulfonation. Generation of electrophilic SO_3 (sulfur trioxide) to attack nucleophilic benzene.

$$C_6H_6 + HOSO_3H \xrightarrow{SO_3} C_6H_5SO_3H + H_2O$$

Sulfuric   Benzenesulfonic
acid   acid

Halogenation. Reaction of Cl_2 or Br_2 with benzene in the presence of a Lewis-acid catalyst.

$$C_6H_6 + X_2 \xrightarrow[\text{or } FeCl_3]{AlCl_3} C_6H_5X + HX \qquad X = Cl, Br$$

Halobenzene

Iodination is possible through the use of special reagents, and fluorination is possible through the Balz–Schiemann reaction.

Friedel–Crafts Alkylation. Reaction of benzene and alkyl halides in the presence of Lewis acids.

$$C_6H_6 + RCl \xrightarrow{AlCl_3} C_6H_5R + HCl$$

Alkylbenzene

Friedel-Crafts Acylation. Reaction of benzene with acyl (carboxylic acid) halides in the presence of anhydrous aluminum chloride.

$$C_6H_6 + COCl \xrightarrow{AlCl_3} C_6H_5COR + HCl$$

Ketone

ADDITION REACTIONS OF BENZENE

Hydrogenation

$$C_6H_6 + 3H_2 \xrightarrow[\text{or Ni at } 180°C]{\text{Pt at R.T.}} C_6H_{12}$$

Cyclohexane

Bromination

$$C_6H_6 + 3Br_2 \xrightarrow[\text{no catalyst}]{\text{Sunlight}} C_6H_6Br_6$$

1,2,3,4,5,6-
hexabromocyclohexane

OXIDATION REACTIONS OF BENZENE

Vigorous Reagents

$$C_6H_6 + \text{Vig. Oxi.} \rightarrow CO_2, H_2O, HCOOH, \text{etc.}$$

Formic acid

Burning

$$2C_6H_6 + (15-n)O_2 \rightarrow (12-n)CO_2 + 6H_2O + nC$$

SUBSTITUTION REACTIONS OF NAPHTHALENE

$$C_{10}H_7 \boxed{H + HO} NO_2 \xrightarrow[\text{H}_2\text{SO}_4 \text{ conc.}]{\text{HNO}_3 \text{ conc.}} C_{10}H_7NO_2 + H_2O$$

(at 50°C)

$$C_{10}H_7 \boxed{H + HO} SO_2OH \xrightarrow{\text{H}_2\text{SO}_4 \text{ conc.}} C_{10}H_7SO_2OH + H_2O$$

(at 80°C)

$$C_{10}H_7 \boxed{H + X} X \xrightarrow[\text{temp}]{\text{boiling}} C_{10}H_7X + HX \quad X = Cl, Br$$

ADDITION REACTIONS OF NAPHTHALENE

$$C_{10}H_8 + 2Na + 2C_2H_5OH \xrightarrow{\text{boiling}} C_{10}H_{10}(1,4)$$

$$C_{10}H_8 + 4Na + 4C_5H_9OH \xrightarrow{\text{boiling}} C_{10}H_{12}(1,2,3,4)$$

$$C_{10}H_8 + 2H_2 + Ni, \text{ Powder} \xrightarrow[200°C]{180°C} C_{10}H_{12}(1,2,3,4)$$

$$C_{10}H_8 + 5H_2 \xrightarrow[\text{catalyst}]{\text{heat}} C_{10}H_{18}$$

$$C_{10}H_8 + Cl_2 \xrightarrow[\text{HCl},20°C]{KClO_3} C_{10}H_8Cl_2 \quad (1,4)$$

$$C_{10}H_8 + 2Cl_2 \xrightarrow[KClO_3, HCl, 20°C]{} C_{10}H_8Cl_4 \quad (1,2,3,4)$$

$$C_{10}H_8 + 2Br_2 \xrightarrow{20°C} C_{10}H_8Br_4 \quad (1,2,3,4)$$

OXIDATION REACTION OF NAPHTHALENE

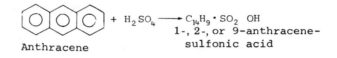 + $H_2SO_4 \longrightarrow C_{14}H_9 \cdot SO_2$ OH
1-, 2-, or 9-anthracene-
Anthracene   sulfonic acid

OXIDATION REACTION OF ANTHRACENE

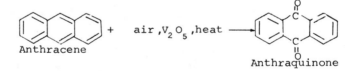
Anthracene + air, V_2O_5, heat $\longrightarrow$

Anthraquinone

The reactions of phenanthrene resemble those of anthracene, since they are isomeric.

8.6 ELECTROPHILIC AROMATIC SUBSTITUTION

The benzene ring may be considered an electron-rich system because of its π electrons: thus it is a nucleophile.

The π electrons of benzene enable it to form new bonds with electron-deficient groups called electrophiles.

A disubstituted benzene ring may have one, two, or three substitution products depending on the orientation of the substituents.

The group attached to the benzene ring determines the orientation of other electrophiles and the reactivity of the benzene ring toward substitution.

A) Activating (electron-donating) groups make the substituted benzene ring more reactive than the unsubstituted benzene.

B) Deactivating (electron-withdrawing) groups make the substituted benzene ring less reactive than the unsubstituted benzene.

C) An ortho, para director is a group that causes attack to occur mainly at positions ortho and para to the group.

D) A meta director is a group that causes attack to occur mainly at position meta to the group.

CLASSIFICATION OF SUBSTITUENT GROUPS

Effect of Groups on Electrophilic Aromatic Substitution

Activating: Ortho,para Directors

Strongly activating
—NH_2(—NHR,—NR_2)
—OH

Moderately activating
—OCH_3(—OC_2H_5,etc.)
—$NHCOCH_3$

Weakly activating
—C_6H_5
—CH_3(—C_2H_5,etc.)

Deactivating: Meta Directors
—NO_2
—$N(CH_3)_3{}^+$
—CN
—COOH(—COOR)
—SO_3H
—CHO,—COR

Deactivating: Ortho,para Directors
—F,—Cl,—Br,—I

Ortho, para directors are activating groups, with the exception of the halogens, which are deactivating groups.

Meta directors are deactivating groups.

Rules for Predicting Orientation in Disubstituted Benzenes

A) If the groups reinforce each other, there is no problem.

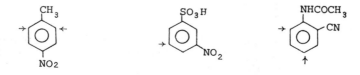

   The arrows indicate the positions where electrophiles can be added.

B) If an o,p-director and m-director are not reinforcing, the o,p-director controls the orientation. The incoming group goes mainly ortho to the m-director.

C) A strongly-activating group, competing with a weakly-activating group, controls the orientation.

D) When two weakly-activating (deactivating) groups or two strongly-activating (deactivating) groups compete, substantial amounts of both isomers are obtained; there is little preference.

E) Very little substitution occurs in the sterically-hindered position between two substituents.

8.7 USES OF BENZENE, NAPTHALENE, AND ANTHRACENE

A) Benzene

   a) Excellent solvent

   b) Used in the blending of some motor fuels

   c) Used in organic synthesis

B) Napthalene

   a) Used in mothballs

   b) Used in fumigation of greenhouses

   c) Used in the manufacture of various dyes and intermediates

   d) Used as raw material in the synthesis of pthalic anhydride

   e) Used as a carbon remover in motor fuels

   f) Tetrahydronapthalene is used as a solvent for fats and waxes

   g) Used in the synthesis of anthraquinone

C) Anthracene

   a) Used in the manufacture of dyes

   b) Formerly used in the synthesis of anthraquinone

CHAPTER 9

ARYL HALIDES

Aryl halides are compounds containing halogens attached directly to a benzene ring. The structural formula is ArX, where the aryl group, Ar, represents phenyl, napthyl, etc., and their derivatives.

9.1 NOMENCLATURE

Aryl halides are named by prefixing the name of the halide to the name of the aryl group. The terms meta, ortho, and para are used to indicate the positions of substituents on a disubstituted benzene ring. Numbers are also used to indicate the positions of the substitutents on a benzene ring.

         Cl Cl

Flouro-benzene   1-Chloronaph-thalene   1-Bromo-2,4-Dichloro-benzene   Ortho-Dichloro-benzene

9.2 PHYSICAL PROPERTIES OF ARYL HALIDES

A) Aryl halides are colorless liquids.

B) Aryl halides are insoluble in water but soluble in organic solvents.

C) Isomeric dihalobenzenes have similar boiling points.

D) Para isomers have a melting point substantially higher than ortho and meta isomers.

E) Because of strong intracrystalline forces, the higher melting para isomer is less soluble in a given solvent than the ortho isomer. It is because of this that purification of the para isomer is often done by recrystallization.

9.3 PREPARATION OF ARYL HALIDES

HALOGENATION BY SUBSTITUTION

$$ArH + X_2 \rightarrow ArX + HX \qquad X_2 = Cl_2, Br_2$$

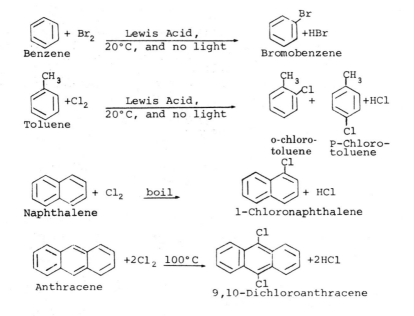

HALOGENATION BY ADDITION

Reaction of phosphorus pentachloride with benzyl alcohol or benzaldehyde to give benzyl chloride or benzal chloride.

67

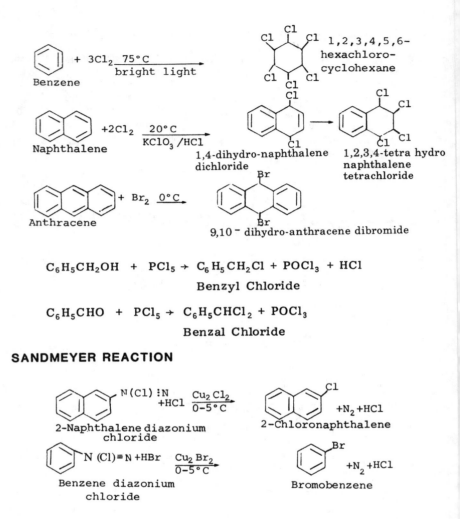

Benzene + 3Cl$_2$ $\xrightarrow[\text{bright light}]{75°C}$ 1,2,3,4,5,6-hexachloro-cyclohexane

Naphthalene + 2Cl$_2$ $\xrightarrow[\text{KClO}_3/\text{HCl}]{20°C}$ 1,4-dihydro-naphthalene dichloride → 1,2,3,4-tetra hydro naphthalene tetrachloride

Anthracene + Br$_2$ $\xrightarrow{0°C}$ 9,10 - dihydro-anthracene dibromide

$$C_6H_5CH_2OH + PCl_5 \rightarrow C_6H_5CH_2Cl + POCl_3 + HCl$$
Benzyl Chloride

$$C_6H_5CHO + PCl_5 \rightarrow C_6H_5CHCl_2 + POCl_3$$
Benzal Chloride

SANDMEYER REACTION

2-Naphthalene diazonium chloride $\text{N(Cl)}:\text{N} + \text{HCl}$ $\xrightarrow[\text{0-5°C}]{\text{Cu}_2\text{Cl}_2}$ 2-Chloronaphthalene $+N_2 + HCl$

Benzene diazonium chloride $\text{N (Cl)} \equiv \text{N} + \text{HBr}$ $\xrightarrow[\text{0-5°C}]{\text{Cu}_2\text{Br}_2}$ Bromobenzene $+N_2 + HCl$

REPLACEMENT OF THE NITROGEN OF A DIAZONIUM SALT

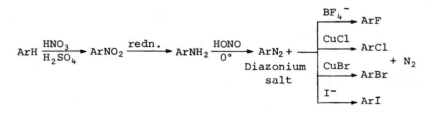

$$\text{ArH} \xrightarrow[\text{H}_2\text{SO}_4]{\text{HNO}_3} \text{ArNO}_2 \xrightarrow{\text{redn.}} \text{ArNH}_2 \xrightarrow[0°]{\text{HONO}} \text{ArN}_2 + \text{Diazonium salt}$$

- BF$_4^-$ → ArF
- CuCl → ArCl
- CuBr → ArBr + N$_2$
- I$^-$ → ArI

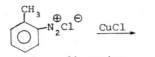

o-Toluenediazonium
chloride

o-Chlorotoluene

TREATMENT OF ARYLTHALLIUM COMPOUNDS WITH IODIDE

$$\left[ArH + Tl(OOCCF_3)_3 \longrightarrow\right] \underset{\substack{Arylthallium \\ trifluoroacetate}}{ArTl(OOCCF_3)_2} + KI \longrightarrow \underset{\substack{For \ iodides \\ only}}{ArI}$$

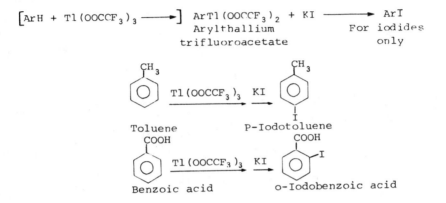

Toluene

P-Iodotoluene

Benzoic acid

o-Iodobenzoic acid

9.4 REACTION OF ARYL HALIDES

FORMATION OF GRIGNARD REAGENT

$$ArBr + Mg \xrightarrow{dry \ ether} ArMgBr$$

$$ArCl + Mg \xrightarrow{\substack{tetrahydro- \\ furan}} ArMgCl$$

Substitution in the ring (electrophilic aromatic substitution). X: Deactivates and directs ortho, para in electrophilic aromatic substitution.

Nucleophilic aromatic substitution (bimolecular displacement).

$$Ar:X + :B^{\ominus} \rightarrow Ar:B + :X^{\ominus}$$

Ar must contain strongly electron-withdrawing groups ortho and/or para to it.

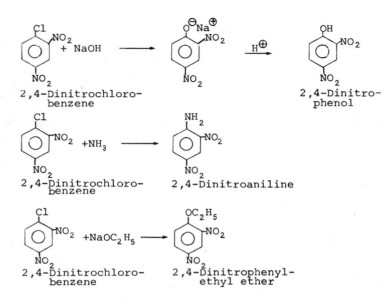

2,4-Dinitrochloro-benzene → 2,4-Dinitro-phenol

2,4-Dinitrochloro-benzene +NH₃ → 2,4-Dinitroaniline

2,4-Dinitrochloro-benzene +NaOC₂H₅ → 2,4-Dinitrophenyl-ethyl ether

REPLACEMENT OF THE HALOGEN ATOM

$$C_6H_5X + HNH_2 \text{ (excess, aq.)} \xrightarrow[\text{pressure}]{\text{heat}} C_6H_5NH_2 + HX$$

$$C_6H_5X + NaOH \text{ aq.} \xrightarrow[\text{pressure}]{\text{heat}} C_6H_5OH + NaX$$

$$C_6H_5X + 2Na + RX \rightarrow C_6H_5R + 2NaX$$

$$C_6H_5X + XMgR \rightarrow C_6H_5R + MgX_2$$

REPLACEMENT OF THE HYDROGEN ATOM

$$X - C_6H_4H + X \cdot X \xrightarrow[\text{catalyst}]{\text{diffused light, 20°C}} X - C_6H_4 - X + HX$$

$$X - C_6H_4H + HONO_2 \xrightarrow[\substack{H_2SO_4,\text{Conc.} \\ 50-60°C}]{HNO_3,\text{Conc.}} X - C_6H_4 - NO_2 + H_2O$$

$$X - C_6H_4 - H + HOSO_2OH \xrightarrow{H_2SO_4/SO_3} X - C_6H_4 - SO_2OH + H_2O$$

Nucleophilic Aromatic substitution. Elimination-addition (Benzyne) mechanism:

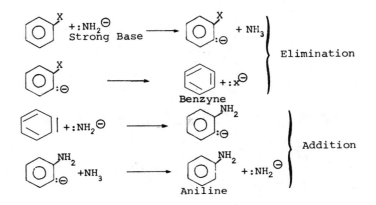

Elimination

Addition

Benzyne

Aniline

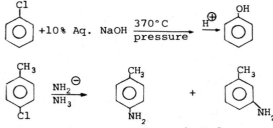

p-Chlorotoluene   p-and m-Aminotoluene

CHAPTER 10

ETHERS AND EPOXIDES

Ethers are hydrocarbon derivatives in which two alkyl or aryl groups are attached to an oxygen atom. The structural formula of an ether is R-O-R', where R and R' may or may not be the same.

ETHERS

10.1 STRUCTURE

Ethers and alcohols are metameric. They are functional isomers of alcohols with the same elemental composition

$$CH_3OCH_3 \quad \text{and} \quad CH_3CH_2OH$$

10.2 NOMENCLATURE (IUPAC SYSTEM)

COMMON NAMES

The attached groups are named in alphabetical order, followed by the word ether.

$CH_3CH_2-O-CH_2CH_2CH_3$
Ethyl propyl ether

CH_3-O-

Methyl phenyl ether

For symmetrical ethers (having the same groups), the compound is named using either the name of the group or the prefix "Di-."

Ex. CH_3-O-CH_3

Methyl ether or
Dimethyl ether

In the IUPAC system, ethers are named as alkoxyalkanes. The larger alkyl group is chosen as the stem.

Ex.

$$CH_3-\overset{\overset{\displaystyle Cl}{|}}{\underset{\underset{\displaystyle Cl}{|}}{C}} - \overset{}{\underset{\underset{\displaystyle OCH_2CH_3}{|}}{CH}} - CH_3$$

3,3-dichloro-2-ethoxybutane

10.3 PHYSICAL AND CHEMICAL PROPERTIES OF ETHERS

A) Ethers are much more volatile than isomeric alcohols.

B) Ethyl ether is colorless, highly volatile, flammable, less dense than and only partially soluble in water; it has a characteristic odor, and is an excellent solvent.

C) Ethers are fairly unreactive to many reagents.

D) Tetrahydrofuran (THF), a cyclic ether, has a boiling point of 67°C, is an important solvent, and is miscible with water.

E) Ethers undergo cleavage in the presence of strong acids.

F) Volatility, flammability, and solubility in water decreases with an increase in C-content. The densities show a gradual increase with increasing molecular weight.

10.4 PREPARATION OF ETHERS

WILLIAMSON REACTION

$RX + NaOR' \rightarrow ROR' + NaX$ R' = alkyl or aryl

alkoxide

73

Ex. CH_3CH_2I + $NaOCH_2CH_3$ → $CH_3CH_2OCH_2CH_3$ + NaI

    ethyl iodide sodium   ethyl ether

              ethoxide

REACTION OF AN ALKYL HALIDE SILVER OXIDE

$$RX + Ag_2O + XR' \xrightarrow{\text{heat}} ROR' + 2AgX$$

Ex. $2CH_3I$ + $Ag_2O \xrightarrow{\text{heat}} CH_3OCH_3$ + $2AgI$

    methyl iodide   methyl ether

DEHYDRATION OF ALCOHOLS

a) $ROH + HOSO_2OH \xrightarrow{\text{cold}} ROSO_2OH + H_2O$

   $ROSO_2OH + HOR' \xrightarrow{\text{heat}} ROR' + H_2SO_4$

Ex. $CH_3CH_2OH + HOSO_2OH \xrightarrow{\text{cold}} CH_3CH_2OSO_2OH$

                                      + H_2O

    ethanol   ethyl hydrogen

                              sulfate

$CH_3CH_2OSO_2OH + HOCH_2CH_3 \xrightarrow{\text{heat}} CH_3CH_2OCH_2CH_3$

                                      + H_2SO_4

                            ethyl ether

b) $2ROH \xrightarrow[\text{240-260°C}]{Al_2O_3} ROR + H_2O$

Ex. $2CH_3CH_2OH \xrightarrow[\text{240-250°C}]{Al_2O_3} CH_3CH_2OCH_2CH_3 + H_2O$

    ethyl alcohol   ethyl ether

ALKOXYMERCURATION–DEMERCURATION

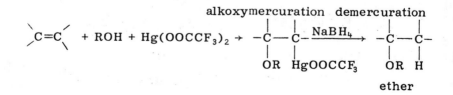

ether

Ex.

$$
\begin{array}{c}
\text{CH}_3 \\
| \\
\text{CH}_3\text{-C-CH=CH}_2 \\
| \\
\text{CH}_3
\end{array}
+ \text{CH}_3\text{CH}_2\text{OH} \xrightarrow{\text{Hg(OOCCF}_3)_2} \xrightarrow{\text{NaBH}_4}
$$

3,3-dimethyl-1-butene

$$
\begin{array}{c}
\text{CH}_3 \\
| \\
\text{CH}_3\text{-C}\text{---}\text{CH-CH}_3 \\
| \quad | \\
\text{CH}_3 \quad \text{OC}_2\text{H}_5
\end{array}
$$

3,3-dimethyl-2-ethoxybutane

10.5 REACTIONS OF ETHERS

SINGLE CLEAVAGE AT THE OXYGEN LINKAGE

a)   HI, Cold and Conc.

   $R-\boxed{O-R'+H}I \rightarrow R'-OH + R-I$

b)   Sulfuric acid, conc., heat

   $R \boxed{O-R'+H} O-SO_2-OH \rightarrow R'-OH + R-O-SO_2-OH$

c)   Steam under pressure, 150°C

   $R-\boxed{O-R'+H}OH \rightarrow R-OH + R'-OH$

DOUBLE CLEAVAGE

a)   Phosphorus pentachloride, heat

   $R-O-R + PCl_5 \rightarrow 2R-Cl + POCl_3$

b)   HI, conc., heat

   $R-O-R + 2HI \rightarrow 2R-I + H_2O$

c)   Sulfuric acid, heat

   $R-O-R + 2HO-SO_2-OH \xrightarrow{\text{heat}} H_2O + 2R-O-SO_2-OH$

$$R-HC \boxed{H + X} - X \rightarrow R-HC-X + HX \quad X = Cl,Br$$
$$\quad\ \ | \qquad\qquad\qquad\qquad |$$
$$\quad R-O \qquad\qquad\qquad\ \ R-O$$

Ex. $CH_3-CH_2-O-CH_2-CH_3 + Cl_2 \xrightarrow{dark} CH_3CH\ Cl-O-CH_2CH_3$

$$+ HCl$$

$$CH_3-CH_2-O-CH_2-CH_3 + 10Cl_2 \xrightarrow{light} CCl_3-CCl_2-O-CCl_2-CCl_3$$

$$+ 10HCl$$

EPOXIDES
10.6 STRUCTURE

Epoxides are cyclic ethers in which the oxygen is included in a three-membered ring.

An epoxide:
Ethylene oxide

10.7 PREPARATION OF EPOXIDES

Oxidation of ethylene by air (oxygen) on a silver catalyst

$$2CH_2 = CH_2 + O_2 \xrightarrow[290°C]{260-} 2CH_2-CH_2$$
$$\hspace{7cm}\diagdown\diagup$$
$$\hspace{6.8cm}O$$

ethylene   Ag catalyst   ethylene oxide

OXIDATION OF ALKENES WITH PEROXYACIDS

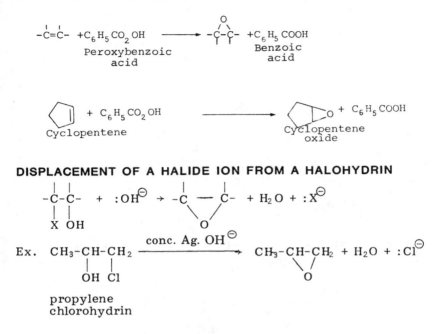

DISPLACEMENT OF A HALIDE ION FROM A HALOHYDRIN

$$-\overset{|}{\underset{\underset{X}{|}}{C}}-\overset{|}{\underset{\underset{OH}{|}}{C}}- \ + \ :OH^{\ominus} \ \rightarrow \ -\overset{|}{\underset{\diagdown}{C}}\underset{O}{}\overset{|}{\underset{\diagup}{C}}- \ + \ H_2O \ + \ :X^{\ominus}$$

Ex. $CH_3-\underset{\underset{OH}{|}}{CH}-\underset{\underset{Cl}{|}}{CH_2} \quad\xrightarrow{\text{conc. Ag. OH}^{\ominus}}\quad CH_3-CH\underset{\diagdown O\diagup}{}CH_2 \ + \ H_2O \ + \ :Cl^{\ominus}$

propylene
chlorohydrin

10.8 REACTIONS OF EPOXIDES

ACID–CATALYZED CLEAVAGE

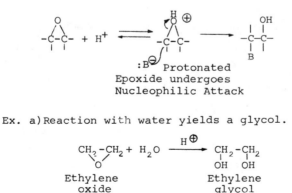

Protonated
Epoxide undergoes
Nucleophilic Attack

Ex. a) Reaction with water yields a glycol.

$$\underset{\underset{O}{\diagdown\diagup}}{CH_2-CH_2} \ + \ H_2O \ \xrightarrow{\ H^{\oplus}\ } \ \underset{\underset{OH}{|}\quad\underset{OH}{|}}{CH_2-CH_2}$$

Ethylene   Ethylene
oxide   glycol

b) Reaction with an alcohol yields a compound that is both an ether and an alcohol.

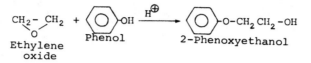

BASE-CATALYZED CLEAVAGE

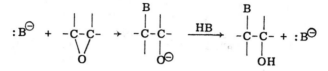

epoxide undergoes
nucleophilic attack

Ex. $C_2H_5O^- Na^+$ + $CH_2—CH_2$ → $C_2H_5OCH_2CH_2OH$

sodium   ethylene   2-ethoxyethanol
ethoxide   oxide

NH_3 + $CH_2—CH_2$ → $H_2NCH_2CH_2OH$

ethylene   2-aminoethanol
oxide

REACTION WITH GRIGNARD REAGENTS

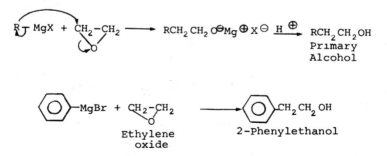

78

CHAPTER 11

ALCOHOLS AND GLYCOLS

Alcohols are hydrocarbon derivatives in which one or more hydrogen atoms have been replaced by the $-OH$ (hydroxyl) group. They have the general formula $R-OH$, where R may be either alkyl or aryl.

11.1 NOMENCLATURE (IUPAC SYSTEM)

Alcohols are named by replacing the "-e" ending of the corresponding alkane with the suffix "ol." The alcohol may also be named by adding the name of the R group to the same alcohol.

Ex. CH_3CH_2OH Ethanol or ethyl alcohol

Depending on what carbon atom the hydroxyl group is attached to, the alcohol is prefixed as follows:

A) Primary ($-OH$ attached to 1° carbon) alcohols are prefixed "n-" or "1-".

B) Secondary ($-OH$ attached to 2° carbon) alcohols are prefixed "sec-" or "2-".

C) Tertiary ($-OH$ attached to 3° carbon) alcohols are prefixed "tert-" or "3-".

Ex. $CH_3CH_2CH_2CH_2OH$

n- or 1- butanol

$CH_3CH_2CHCH_3$
|
OH

sec- or 2-butanol

$$CH_3CH_2-\overset{\overset{\displaystyle CH_3}{|}}{\underset{\underset{\displaystyle OH}{|}}{C}}-CH_3$$

tert- or 3- pentanol

11.2 PHYSICAL PROPERTIES OF ALCOHOLS

A) Alcohols are high-boiling.

B) Lower alcohols are soluble in water.

C) Refractive index, density, and boiling point of alcohols increase with an increase in C-content, but solubility in water decreases.

D) The introduction of additional -OH groups into a molecule tends to intensify the characteristic properties of the hydroxyl group.

E) In general, for a given number of carbon atoms, the boiling point, the sweetness, and the solubility of an alcohol in water, increase with the number of hydroxyl groups.

11.3 PREPARATION OF MONOHYDROXY ALCOHOLS

A) Substitution of -OH for -X when alkyl halides are treated with aqueous sodium or potassium hydroxide.

$$R-X + NaOH(aq) \rightarrow R-OH + NaX$$

Ex. $CH_3CH_2-Cl + NaOH(aq) \rightarrow CH_3CH_2-OH + NaCl$

chloroethane   ethanol

B) Introduction of alkyl and hydrogen groups into a carbonyl or epoxy compound upon treatment with a Grignard reagent in dry ether and subsequent hydrolysis.

a) $H-CHO + R-Mg-X \rightarrow R-CH_2-O-Mg-X$

$$R-CH_2-O-Mg-X + HX(aq) \rightarrow R-CH_2-OH + MgX_2$$

primary alcohol

Ex.

$$\underset{\underset{H}{|}}{\overset{\overset{H}{|}}{C}}=O + CH_3-Mg-I \rightarrow CH_3-\underset{\underset{H}{|}}{\overset{\overset{H}{|}}{C}}-O-Mg-I$$

formaldehyde

$$CH_3-CH_2-O-Mg-I + HCl(aq) \rightarrow CH_3-CH_2-OH + Mg-I-Cl$$

b) 1. $R-CHO + R'-Mg\,X + HX(aq) \rightarrow R-CHOH-R' + MgX_2$

secondary alcohol

2. $R\overset{\overset{\displaystyle O}{\diagup\,\diagdown}}{CH}-CH_2 + R'-Mg-X + HX(aq) \rightarrow R-CHOH-CH_2-R'$

$+ MgX_2$

c) 1. $R_2C = O + R'MgX + 2HX(aq) \rightarrow R_2R'C-OH + MgX_2$

tertiary
alcohol

2. $R_2\overset{\overset{\displaystyle O}{\diagup\,\diagdown}}{C}-CH_2 + R_2'Mg + HX(aq) \rightarrow R_2C(OH)-CH_2-R'$

$+ MgX_2 + R'H$

C) By the catalytic hydrogenation or by the reduction of aldehydes or ketones in acid solution.

a) $R-CHO + Zn + 2H_2O$, acidic $\rightarrow R-CH_2-OH + Zn^{++} + 2OH^-$

aldehyde   1° alcohol

Ex. $CH_3-CH_2-CHO + Zn + 2H_2O$, acidic $\rightarrow CH_3CH_2-CH_2-OH$

$+ Zn^{++} + 2OH^-$

1-propanol

b) $R_2C = O + Zn + 2H_2O$, acidic $\rightarrow R_2CHOH + Zn^{++} + 2OH^-$

ketone   2° alcohol

81

D) Hydration of Alkenes

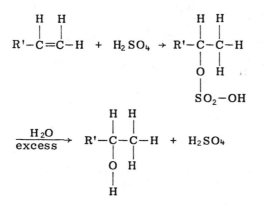

$$R'-\underset{\underset{H}{|}}{\overset{\overset{H}{|}}{C}}-\underset{\underset{H}{|}}{\overset{\overset{H}{|}}{C}}-\underset{\underset{H}{|}}{\overset{\overset{H}{|}}{C}}-R + cracking \rightarrow R'-\overset{\overset{H}{|}}{C}=\overset{\overset{H}{|}}{C}-H + H-\underset{\underset{H}{|}}{\overset{\overset{H}{|}}{C}}-R$$

$$R > R'$$

$$R'-\overset{\overset{H}{|}}{C}=\overset{\overset{H}{|}}{C}-H + H_2SO_4 \rightarrow R'-\overset{\overset{H}{|}}{\underset{\underset{SO_2-OH}{\underset{|}{O}}}{C}}-\overset{\overset{H}{|}}{C}-H$$

$$\xrightarrow[\text{excess}]{H_2O} R'-\underset{\underset{H}{\underset{|}{O}}}{\overset{\overset{H}{|}}{C}}-\overset{\overset{H}{|}}{C}-H + H_2SO_4$$

E) By the action of heat and pressure on a mixture of carbon monoxide or carbon dioxide and hydrogen in the presence of a catalyst (zinc chromite).

$$CO + 2H_2 \xrightarrow[\text{400-500°C ,200 atm.}]{\text{zinc chromite}} CH_3OH$$
$$\text{methanol}$$

F) Reaction of amines with nitrous acid.

a) $CH_3-NH_2 + HO-NO \xrightarrow[H^{\oplus}]{NaNO_2} CH_3OH + N_2 + H_2O$
   Methyl Amine   Methanol

b) $CH_3-CH_2-NH_2 + HO-NO \xrightarrow[H^{\oplus}]{Na-NO_3} CH_3-CH_2-OH + N_2 + H_2O$
   Ethyl Amine   Ethanol

G) Oxymercuration-Demercuration

$$\overset{\diagdown}{\diagup}C=C\overset{\diagup}{\diagdown} + H_2O \xrightarrow{Hg(OAc)_2} \underset{\underset{OH}{|}}{-C}-\underset{\underset{HgOAc}{|}}{C}- \xrightarrow{NaBH_4} \underset{\underset{OH}{|}}{-C}-\underset{\underset{H}{|}}{C}-$$

Markovnikov
addition

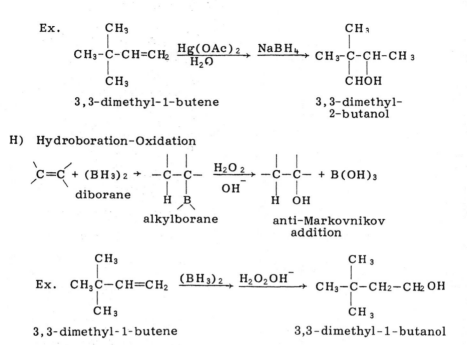

Ex.

$CH_3-\underset{\underset{CH_3}{|}}{\overset{\overset{CH_3}{|}}{C}}-CH=CH_2$ $\xrightarrow[H_2O]{Hg(OAc)_2}$ $\xrightarrow{NaBH_4}$ $CH_3-\underset{\underset{CHOH}{|}}{\overset{\overset{CH_3}{|}}{C}}-CH-CH_3$

3,3-dimethyl-1-butene   3,3-dimethyl-2-butanol

H) Hydroboration-Oxidation

$\underset{/}{\overset{\backslash}{C}}=\underset{\backslash}{\overset{/}{C}}$ + $(BH_3)_2$ → $-\underset{\underset{H}{|}}{C}-\underset{\underset{B}{|}}{C}-$ $\xrightarrow[OH^-]{H_2O_2}$ $-\underset{\underset{H}{|}}{C}-\underset{\underset{OH}{|}}{C}-$ + $B(OH)_3$

diborane   alkylborane   anti-Markovnikov addition

Ex. $CH_3\underset{\underset{CH_3}{|}}{\overset{\overset{CH_3}{|}}{C}}-CH=CH_2$ $\xrightarrow{(BH_3)_2}$ $\xrightarrow{H_2O_2OH^-}$ $CH_3-\underset{\underset{CH_3}{|}}{\overset{\overset{CH_3}{|}}{C}}-CH_2-CH_2OH$

3,3-dimethyl-1-butene   3,3-dimethyl-1-butanol

11.4 REACTIONS OF MONOHYDROXY ALCOHOLS

A) Replacement of the hydrogen atom of the hydroxyl group when treated with:

a) Active metals

$$2R-OH + 2Na \rightarrow 2R-ONa + H_2$$

b) Acid halides

$$R-O\boxed{H + X}OC-R' \rightarrow R-OOC-R' + HX$$

c) Organic acids

$$R-O\boxed{H + HO}OC-R' \xrightarrow[\text{or } P_2O_5]{H_2SO_4} R-OOC-R' + H_2O$$

d) Alkyl hydrogen sulfates

$$R-O\boxed{H + HOSO_2O}\overset{\prime}{R} \rightarrow R-O\overset{\prime}{R} + H_2SO_4$$

e) Grignard reagent

$$R-O\boxed{H + R}-Mg-X \rightarrow R-O-Mg-X + RH$$

B) Replacement of the hydroxyl group when treated with:

a) Hydriodic acid/(red phosphorus)

$$R-\boxed{OH + H}I \rightarrow RI + H_2O$$

b) Hydrobromic acid, dry (or hydrochloric acid)

$$R-\boxed{OH + H}Br \rightarrow R-Br + H_2O$$

c) Hydrochloric acid (or HBr)/conc. sulfuric acid

$$R-\boxed{OH + H}Cl \rightarrow R-Cl + H_2O$$

d) Sulfuric acid

$$R-\boxed{OH + H}OSO_2-OH \rightarrow H_2O + R-O-SO_2-OH$$

e) Nitric acid

$$R-\boxed{OH + H}NO_3 \rightarrow H_2O + R-NO_3$$

f) Phosphorus trihalides

$$3R-OH + P X_3 \rightarrow P(OH)_3 + 3RX$$

g) Phosphorus pentahalides

$$R-OH + PCl_5 \rightarrow R-CL + HCl + POCl_3$$

C) Dehydration of alcohols by acids to give unsaturated derivatives.

$$R-CHOH-CH_2-R + P_2O_5 \rightarrow R-HC=CH-R + 2HPO_3$$

Alkene

D) Oxidation of alcohols, to give derivatives, when treated with:

a) One mole of dichromic acid per three moles of $R-CH_2-OH$.

$$R-CH_2-OH + \text{oxidation} \rightarrow R-CHO \rightarrow R-CO-OH \rightarrow \begin{array}{l}\text{oxidized}\\\text{derivatives}\end{array}$$

$\longrightarrow$ cleavage   aldehyde

Ex. $3CH_3-CH_2-OH + Na_2Cr_2O_7/4H_2SO_4$

$$\rightarrow 3CH_3-CHO + Na_2SO_4 + Cr_2(SO_4)_3 + 7H_2O$$
acetaldehyde

b) Two moles of dichromic acid per three moles of $R-CH_2-OH$.

Ex. $3CH_3-CH_2-OH + 2Na_2Cr_2O_7/8H_2SO_4$

$$\rightarrow 3CH_3-CO-OH + 2Na_2SO_4 + 2Cr_2(SO_4)_3 + 11H_2O$$
Acetic acid

c) One mole of dichromic acid per three moles of $R_2-CH-OH$.

$$R_2-CH-OH + \text{oxidation} \rightarrow R_2C=O \rightarrow \text{oxidized derivative}$$
$$\rightarrow \text{cleavage}$$
ketone

Primary alcohols can be oxidized to form aldehydes and acids.

Ex. $CH_3CH_2CH_2OH + KMnO_4 \rightarrow CH_3CH_2COOH$
n-propanol   propionic acid

$$CH_3CH_2CH_2OH + K_2Cr_2O_7 \rightarrow CH_3CH_2\overset{\displaystyle H}{\underset{|}{C}}=O$$
propionaldehyde

Secondary alcohols can be oxidized to form ketones.

Ex. $CH_3-\overset{\displaystyle CH_3}{\underset{|}{CH}}-CH_2OH \xrightarrow[\text{or } CrO_3]{K_2Cr_2O_7} CH_3-\overset{\displaystyle CH_3}{\underset{|}{C}}-\overset{\displaystyle H}{\underset{|}{C}}=O$
isobutyl alcohol   2-methyl propanol

Tertiary alcohols cannot be oxidized.

11.5 USES OF ALCOHOLS

A) Alcohols are widely used in synthesis, especially in that of ester, and as solvents.

B) Methanol is used as an anti-freeze and in the production of methanal (formaldehyde), which is used in the synthesis of resins.

C) Ethanol is used as a solvent, synthetic intermediate, anti-freeze, and as an ingredient in alcoholic beverages.

D) Butyl and amyl alcohols are used in the preparation of esters for the lacquer industry.

11.6 GLYCOLS

Alcohols containing more than one hydroxyl group (polyhydroxyalcohols) are represented by the general formula $CnH_{2n + 2y}(OH)y$. Polyhydroxyalcohols containing two hydroxyl groups are called glycols or diols.

Ex. 1,3-butanediol

$$CH_3-CH-CH_2-CH_2-OH$$
$$\overset{|}{O}H$$

11.7 PREPARATION OF GLYCOLS

Treatment of ethylene with hypochlorous acid and subsequent hydrolysis.

$H_2C = CH_2 + HO-Cl$, aq. $\rightarrow$ $HO-CH_2-CH_2-Cl + NaHCO_3$,aq
halohydrin
$\rightarrow$ $HO-CH_2-CH_2-OH$ (glycol) + $NaCl + CO_2$

Oxidation of ethylene with the presence of gold or silver and the addition of water.

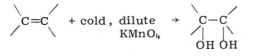

$$H_2C = CH_2 + \tfrac{1}{2}O_2 \xrightarrow{Au/Ag} H_2C\underset{\text{epoxide}}{\overset{O}{\triangle}}CH_2$$

$$\xrightarrow{+H_2O} HO-CH_2-CH_2-OH$$

Cold dilute potassium permanganate causes hydroxylation of alkenes to yield glycols.

$$\underset{/}{\overset{\backslash}{}}C=C\underset{\backslash}{\overset{/}{}} + \underset{KMnO_4}{\text{cold, dilute}} \rightarrow \underset{\underset{OH\ OH}{|\ \ |}}{\overset{\backslash}{}}C-C\overset{/}{}$$

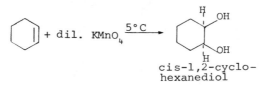

cis-1,2-cyclo-
hexanediol

Aldol condensation of α,α–dialkylacetaldehydes with formaldehyde:

$$R_2CHCHO + CH_2O \xrightarrow{KOH} HOCH_2\overset{\overset{R}{|}}{\underset{\underset{R}{|}}{C}}CH_2OH$$

Reduction of dicarbonyl compounds.

Ex.
$$CH_3\overset{O}{\overset{||}{C}}(CH_2)_3\overset{O}{\overset{||}{C}}CH_3 + NaBH_4 \xrightarrow{C_2H_5OH} CH_3\overset{OH}{\overset{|}{C}}H(CH_2)_3\overset{OH}{\overset{|}{C}}HCH_3$$

2,6-heptanediol

Cleavage of epoxides by water in the presence of mineral acids to give trans-glycol.

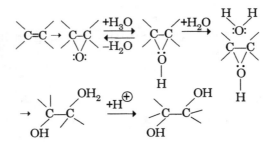

Hydrolysis of alkyl dihalides:

$$R-\underset{\underset{X}{|}}{C}H-\underset{\underset{X}{|}}{C}H-R + 2:OH^{\ominus} \xrightarrow[H_2O]{Na_2CO_3} R-\underset{\underset{OH}{|}}{C}H-\underset{\underset{OH}{|}}{C}H-R + 2:X^{\ominus}$$

where X = Cl, Br

Hydrolysis of halohydrins in the presence of bases.

$$R-\underset{\underset{X}{|}}{C}H-\underset{\underset{OH}{|}}{C}H-R + :OH^{-} \rightarrow R-\underset{\underset{OH}{|}}{C}H-\underset{\underset{OH}{|}}{C}H-R + :X^{\ominus}$$

11.8 REACTION OF GLYCOLS

OXIDATION OF ETHANEDIOL

$$\begin{array}{ccccc}
\underset{\underset{CH_2OH}{|}}{CH_2OH} + \text{oxidation} \rightarrow & \underset{\underset{CH_2OH}{|}}{CHO} \rightarrow & \underset{\underset{CH_2OH}{|}}{COOH} \rightarrow & \underset{\underset{CO-OH}{|}}{CO-OH}
\end{array}$$

ethanediol   hydroxy   hydroxy   oxalic
               ethanol   ethanoic   acid
                          acid

Nitration of Ethanediol

$$HO-CH_2-CH_2-OH + 2HNO_3 \left(\xrightarrow{H_2SO_4} O_2-NO-CH_2-CH_2-ONO_2 \right.$$

$$+ 2H_2O$$

1,2 –dinitroethane

Glycols undergo oxidative cleavage by periodic acid, yielding two carbonyl compounds:

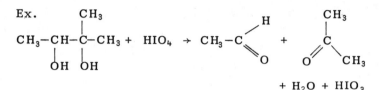

Ex.

$$CH_3-\underset{\underset{OH}{|}}{CH}-\underset{\underset{OH}{|}}{\overset{\overset{CH_3}{|}}{C}}-CH_3 + HIO_4 \rightarrow CH_3-C\overset{H}{\underset{O}{\diagdown}} + \overset{CH_3}{\underset{O\diagdown CH_3}{C}}$$

$$+ H_2O + HIO_3$$

Lead tetraacetate, $Pb(CH_3COO)_4$, cleaves glycols by oxidation.

Ex.

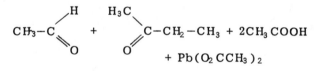

$$CH_3-C\overset{H}{\underset{O}{\diagdown}} + \overset{H_3C\diagdown}{\underset{O\diagup}{C}}-CH_2-CH_3 + 2CH_3COOH$$

$$+ Pb(O_2CCH_3)_2$$

CHAPTER 12

CARBOXYLIC ACIDS

Carboxylic acids contain a carboxyl group

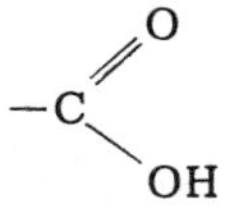

bonded to either an alkyl group (RCOOH) or an aryl group (ArCOOH).

HCOOH is formic acid (methanoic acid)

CH_3COOH is acetic acid (ethanoic acid)

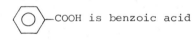 COOH is benzoic acid

12.1 NOMENCLATURE (IUPAC SYSTEM)

The longest chain carrying the carboxyl group is considered the parent structure and is named by replacing the "-e" ending of the corresponding alkane with "-oic acid."

$CH_3CH_2CH_2CH_2COOH$ Pentanoic acid

$CH_3CH_2CHCOOH$
|
CH_3 2-Methylbutanoic acid

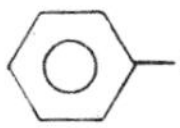

CH_2CH_2COOH 3-Phenylpropanoic acid

$CH_3CH = CHCOOH$ 2-Butenoic acid

90

The position of the substituent is indicated by a number (e.g. $\overset{5}{C}-\overset{4}{C}-\overset{3}{C}-\overset{2}{C}-\overset{1}{COOH}$). The name of a salt of a carboxylic acid consists of the name of the cation followed by the name of the acid with the ending "-ic acid" changed to "-ate."

⬡—COONa	$(CH_3COO)_2 Ca$	$HCOONH_4$
Sodium benzoate	Calcium acetate	Ammonium formate

12.2 PHYSICAL PROPERTIES OF CARBOXYLIC ACIDS

A) Whether the attached group is aliphatic or aromatic, saturated or unsaturated, substituted or unsubstituted, the properties of the carboxyl group are essentially the same.

B) Carboxylic acids are polar molecules and can form hydrogen bonds with each other and other kinds of molecules.

C) The first four members are miscible in water, the 5-carbon acid is partially soluble, and the higher acids are insoluble. This is due to hydrogen bonding.

D) Carboxylic acids have high boiling points which increase with carbon content, because pairs of molecules are held together by two hydrogen bonds.

E) Carboxylic acids are soluble in less polar solvents like ether, alcohol, benzene, etc.

F) The odors of lower aliphatic acids progress from the sharp irritating odors of formic acid and acetic acids to the distinctly unpleasant odors of butyric, valeric and caproic acids. Higher acids have little odor because of their low volatility.

12.3 PREPARATION OF CARBOXYLIC ACIDS

OXIDATION OF PRIMARY ALCOHOLS

$$RCH_2OH \xrightarrow{KMnO_4} RCOOH$$

Ex.

$$CH_3CH_2\overset{\overset{\displaystyle CH_3}{|}}{C}HCH_2OH \xrightarrow{KMnO_4} CH_3CH_2\overset{\overset{\displaystyle CH_3}{|}}{C}HCOOH$$

2-methyl-1-butanol   2-methylbutanoic acid

Oxidation of Alkylbenzenes.

$$Ar-R \xrightarrow[K_2Cr_2O_7]{KMnO_4, or} Ar-COOH$$

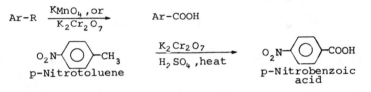

p-Nitrotoluene $\xrightarrow[H_2SO_4, heat]{K_2Cr_2O_7}$ p-Nitrobenzoic acid

Carbonation of Grignard Reagents:

$$RX (or\ ArX) \xrightarrow{Mg} RMgX \xrightarrow[\substack{Carbon-\\ation}]{CO_2} RCOOMgX \xrightarrow{H^+} \substack{RCOOH\\(or\ ArCOOH)}$$

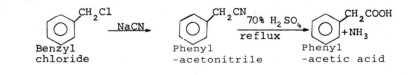

p-Bromo-sec-butyl benzene   p-sec-Butyl benzoic acid

Hydrolysis of nitriles:

$$\substack{R-C\equiv N\\or\\Ar-C\equiv N} +H_2O \xrightarrow[Base]{Acid\ or} \substack{R-COOH\\or\\Ar-COOH} + NH_3$$

Benzyl chloride $\xrightarrow{NaCN}$ Phenyl-acetonitrile $\xrightarrow[reflux]{70\% H_2SO_4}$ Phenyl-acetic acid $+NH_3$

GRIGNARD SYNTHESIS

$$R\!-\!MgX + C \rightarrow RCOO^- Mg X^+ \xrightarrow{H^+} RCOOH + Mg^{++} + 2X^-$$

NITRILE SYNTHESIS

$$R\!-\!X + CN^- \rightarrow R\!-\!C \equiv N + X^-$$

$$RC \equiv N + H_2O \underset{OH^-}{\overset{H^+}{\longrightarrow}} \begin{cases} RCOOH + NH_4^+ \\ RCOO^- + NH_3 \end{cases}$$

Friedel-Craft acylation of aromatic hydrocarbons by acid anhydrides.

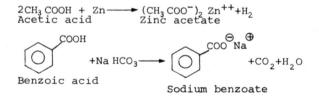

Acid anhydride   Ketocarboxylic acid

12.4 REACTIONS OF CARBOXYLIC ACIDS

ACIDITY SALT FORMATION

$$RCOOH \rightleftharpoons RCOO^- + H^+$$

$$2CH_3COOH + Zn \longrightarrow (CH_3COO^-)_2 Zn^{++} + H_2$$
Acetic acid   Zinc acetate

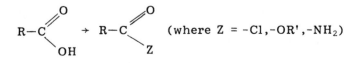

Benzoic acid

Sodium benzoate

CONVERSION INTO FUNCTIONAL DERIVATIVES

$$R\!-\!C\!\!\begin{array}{c} O \\ \diagdown \\ OH \end{array} \rightarrow R\!-\!C\!\!\begin{array}{c} O \\ \diagdown \\ Z \end{array} \quad \text{(where } Z = -Cl, -OR', -NH_2)$$

A) Conversion to acid chlorides:

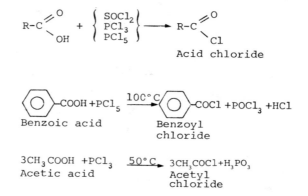

$$R-C \underset{OH}{\overset{O}{\diagup}} + \left\{ \begin{array}{l} SOCl_2 \\ PCl_3 \\ PCl_5 \end{array} \right\} \longrightarrow R-C \underset{Cl}{\overset{O}{\diagup}}$$

Acid chloride

Benzoic acid + PCl_5 $\xrightarrow{100°C}$ Benzoyl chloride $COCl$ + $POCl_3$ + HCl

Benzoic acid → Benzoyl chloride

$3CH_3COOH + PCl_3 \xrightarrow{50°C} 3CH_3COCl + H_3PO_3$

Acetic acid → Acetyl chloride

B) Conversion into esters:

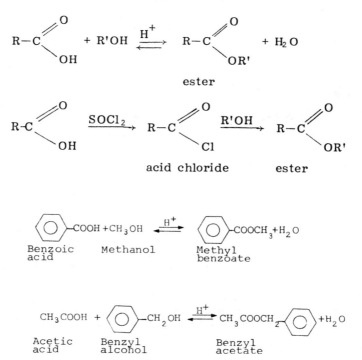

$$R-C \underset{OH}{\overset{O}{\diagup}} + R'OH \underset{\longleftarrow}{\overset{H^+}{\longrightarrow}} R-C \underset{OR'}{\overset{O}{\diagup}} + H_2O$$

ester

$$R-C \underset{OH}{\overset{O}{\diagup}} \xrightarrow{SOCl_2} R-C \underset{Cl}{\overset{O}{\diagup}} \xrightarrow{R'OH} R-C \underset{OR'}{\overset{O}{\diagup}}$$

acid chloride   ester

$COOH + CH_3OH \underset{\longleftarrow}{\overset{H^+}{\longrightarrow}} COOCH_3 + H_2O$

Benzoic acid   Methanol   Methyl benzoate

$CH_3COOH +$ $CH_2OH \underset{\longleftarrow}{\overset{H^+}{\longrightarrow}} CH_3COOCH_2$ $+ H_2O$

Acetic acid   Benzyl alcohol   Benzyl acetate

94

C) Conversion to amides:

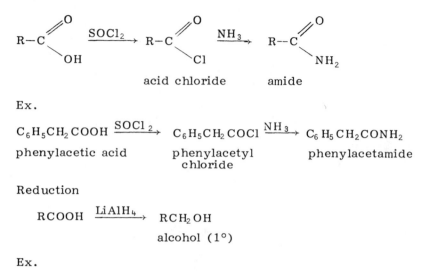

acid chloride   amide

Ex.

$C_6H_5CH_2COOH \xrightarrow{SOCl_2} C_6H_5CH_2COCl \xrightarrow{NH_3} C_6H_5CH_2CONH_2$

phenylacetic acid   phenylacetyl   phenylacetamide
                      chloride

Reduction

$RCOOH \xrightarrow{LiAlH_4} RCH_2OH$

alcohol (1°)

Ex.

$4(CH_3)_3CCOOH + 3LiAlH_4 \xrightarrow{ether} [(CH_3)_3CCH_2O]_4AlLi$

Trimethylacetic
acid $+ 2LiAlO_2 + 4H_2$

$\xrightarrow{H^+} (CH_3)_3CCH_2OH$

2,2-dimethyl-1-propanol

Substitution in alkyl or aryl group:

a) Alpha-halogenation of aliphatic acids
   (Hell-Volhard-Zelinsky reaction)

$R-CH_2COOH + X_2 \xrightarrow{P} \underset{\underset{X}{|}}{R}CHCOOH + HX \quad (X = Cl, Br)$

α-haloacid

Ex.

$CH_3COOH \xrightarrow{Cl_2, P} ClCH_2COOH \xrightarrow{Cl_2, P}$

acetic acid   chloroacetic
               acid

95

$$\rightarrow \ Cl_2CHCOOH \ \rightarrow \ Cl_3CCOOH$$

dichloroacetic   trichloroacetic
acid   acid

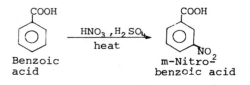

$$CH_3\overset{\underset{\displaystyle CH_3}{|}}{C}HCH_2COOH \xrightarrow{Br_2, P} CH_3-\overset{\underset{\displaystyle}{|}}{C}H-\overset{\underset{\displaystyle Br}{|}}{C}H-COOH$$

isovaleric acid   α-bromoisovaleric acid

b) Ring substitution in aromatic acids

-COOH: deactivates, and directs meta in electrophilic substitution.

$$\underset{\text{Benzoic acid}}{\text{COOH}} \xrightarrow[\text{heat}]{HNO_3, H_2SO_4} \underset{\text{m-Nitro-benzoic acid}}{\text{COOH} \ NO_2}$$

12.5 ACIDITY OF CARBOXYLIC ACIDS

The acidity of carboxylic acid is due to the powerful resonance-stabilization of its anion. This stabilization and resulting acidity are possible only because of the presence of the carbonyl group. The stabilization can be seen in Fig. 12.1.

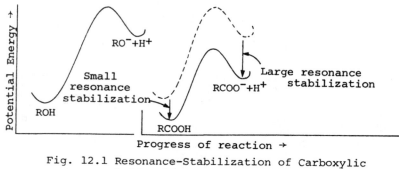

Fig. 12.1 Resonance-Stabilization of Carboxylic Acid Anion

12.6 STRUCTURE OF CARBOXYLATE IONS

A carboxylate ion, according to the resonance theory, is a hybrid of two structures which, being of equal stability, contribute equally. Carbon is joined to each oxygen by a "one and one-half" bond.

IONIZATION OF CARBOXYLIC ACIDS: ACIDITY CONSTANTS

Carboxylic acids are acidic in aqueous solutions. The ionization of carboxylic acids proceeds as follows:

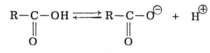

carboxylic   carboxylate
    acid   anion

The ionization of a carboxylic acid produces a carboxylate anion $R-COO^{\ominus}$, which is capable of resonance delocalization of charge.

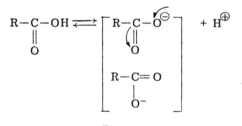

Resonance
delocalization

Electron-withdrawing substituents stabilize the anion and increase acidity.

Electron-releasing substituents destabilize the anion and decrease acidity.

The electron-withdrawing halogens strengthen acids. For example, in order of increasing acidity:

Acetic Acid < Chloroacetic Acid < Dichloroacetic Acid

< Trichloroacetic Acid

ACIDITY CONSTANT

$$RCOOH + H_2O \rightleftharpoons RCOO^- + H_3O^+$$

$$K_a = \frac{[RCOO^-][H_3O^+]}{[RCOOH]}$$

K_a is the acidity constant.

K_a, which is a characteristic of every carboxylic acid, indicates the strength of the acid. As the value K_a increases, the extent of ionization and the strength of the acid increase.

Relative acidities

$$RCOOH > HOH > ROH > HC \equiv CH > NH_3 > RH$$

Relative basicities

$$RCOO^- < HO^- < RO^- < HC \equiv C^- < NH_2^- < R^-$$

CHAPTER 13

CARBOXYLIC ACID DERIVATIVES

Carboxylic acid derivatives are compounds in which the hydroxyl group has been replaced by -Cl, -OOCR, $-NH_2$, or -OR'. These derivatives are called acid chlorides, anhydrides, amides, and esters, respectively.

13.1 ACID CHLORIDES

NOMENCLATURE (IUPAC SYSTEM)

When naming acid chlorides, the ending "-ic acid" in the carboxylic acid is replaced by the ending "-yl chloride."

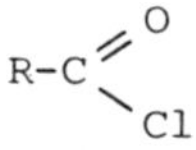

$$R-C \overset{O}{\underset{Cl}{\diagup}}$$

Acid
chloride

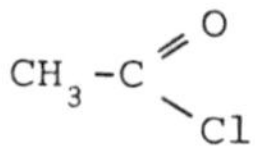

$$CH_3-C \overset{O}{\underset{Cl}{\diagup}}$$

Ethanoyl
chloride
(Acetyl chloride)

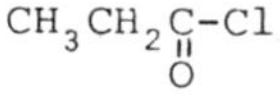

Benzoyl
chloride

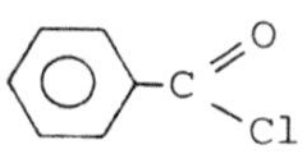

$$CH_3CH_2\underset{O}{\overset{\|}{C}}-Cl$$

Propanoyl
chloride

$$CH_3\underset{CH_3}{\overset{|}{C}H_2}CH_2CH_2\underset{O}{\overset{\|}{C}}-Cl$$

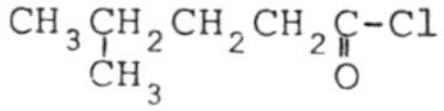

4-Methyl pentanoyl
chloride

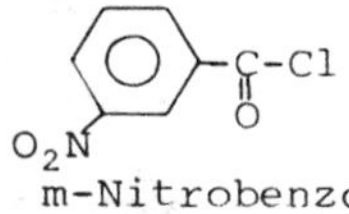

O_2N

m-Nitrobenzoyl
chloride

$$CH_3CH=CHCOCl$$

2-Butenoyl
chloride

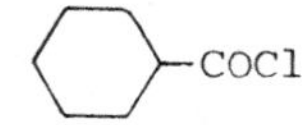

—COCl

Cyclohexane
carbonyl chloride

PHYSICAL PROPERTIES OF ACID DERIVATIVES

The presence of the C=O group makes the acid derivatives polar compounds. Acid chlorides, anhydrides and esters have boiling points that are about the same as those of aldehydes and ketones of comparable molecular weight. The acid derivatives are soluble in the usual organic solvents.

Acid chlorides have sharp, irritating odors, partly because they are readily hydrolyzed to HCl and carboxylic acid. It is becuase of this that they must be protected from moisture.

Acid chlorides are used as intermediates in synthesis. They are fairly reactive compounds and are usually liquids.

PREPARATION OF ACID CHLORIDES

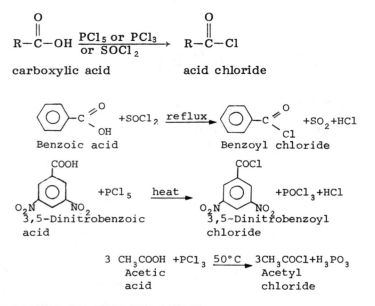

$$\underset{\text{carboxylic acid}}{R-\overset{\overset{\textstyle O}{\|}}{C}-OH} \xrightarrow{\underset{\text{or } SOCl_2}{PCl_5 \text{ or } PCl_3}} \underset{\text{acid chloride}}{R-\overset{\overset{\textstyle O}{\|}}{C}-Cl}$$

Benzoic acid +SOCl₂ reflux → Benzoyl chloride +SO₂+HCl

3,5-Dinitrobenzoic acid +PCl₅ heat → 3,5-Dinitrobenzoyl chloride +POCl₃+HCl

3 CH₃COOH +PCl₃ 50°C → 3CH₃COCl+H₃PO₃
Acetic acid → Acetyl chloride

REACTIONS OF ACID CHLORIDES

A) Conversion into acids and derivatives

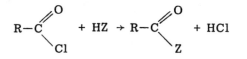

$$R-\overset{\overset{\textstyle O}{\diagup}}{\underset{\diagdown Cl}{C}} + HZ \rightarrow R-\overset{\overset{\textstyle O}{\diagup}}{\underset{\diagdown Z}{C}} + HCl$$

100

a) Conversion into acids. Hydrolysis.

$$RCOCl + H_2O \rightarrow RCOOH + HCl$$
$$\text{an acid}$$

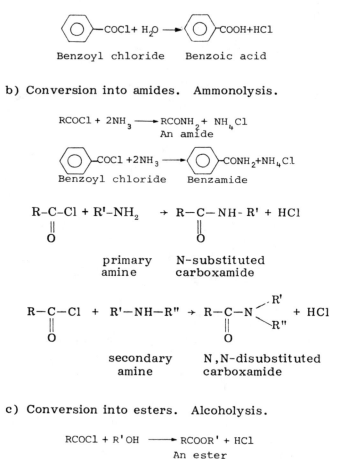

Benzoyl chloride   Benzoic acid

b) Conversion into amides. Ammonolysis.

$$RCOCl + 2NH_3 \longrightarrow RCONH_2 + NH_4Cl$$
$$\text{An amide}$$

Benzoyl chloride   Benzamide

$$R\text{-}C\text{-}Cl + R'\text{-}NH_2 \rightarrow R\text{-}C\text{-}NH\text{-}R' + HCl$$

    primary   N-substituted
    amine   carboxamide

$$R\text{-}C\text{-}Cl + R'\text{-}NH\text{-}R'' \rightarrow R\text{-}C\text{-}N\begin{smallmatrix}R'\\R''\end{smallmatrix} + HCl$$

    secondary   N,N-disubstituted
    amine   carboxamide

c) Conversion into esters. Alcoholysis.

$$RCOCl + R'OH \longrightarrow RCOOR' + HCl$$
$$\text{An ester}$$

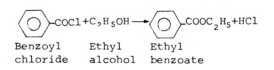

Benzoyl   Ethyl   Ethyl
chloride   alcohol   benzoate

B) Formation of ketones. Friedel-Crafts acylation.

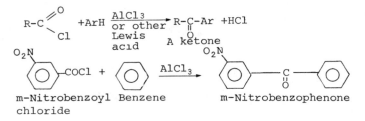

C) Formation of ketones. Reaction with organocadmium compounds.

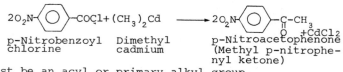

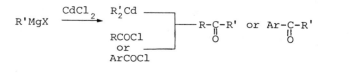

R' must be an acyl or primary alkyl group.

D) Formation of aldehydes by reduction.

a)   RCOCl or ArCOCl $\xrightarrow{\text{LiAlH(OBu-t)}_3}$ RCHO or ArCHO
                                                          Aldehyde

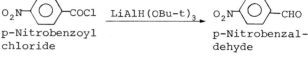

b) Rosenmund reduction of acid chlorides.

$$R-\underset{\underset{O}{\|}}{C}-Cl \;+\; H_2 \xrightarrow{\text{Pd/BaSO}_4} RCHO + HCl$$

                                          aldehyde

E) Reduction to alcohols.

$$2CH_3COCl + LiAlH_4 \rightarrow LiAlCl_2(COCH_2CH_3)_2 \xrightarrow{\overset{+}{H}} 2CH_3CH_2OH$$

102

13.2 CARBOXYLIC ACID ANHYDRIDES

NOMENCLATURE (IUPAC SYSTEM)

When naming acid anhydrides, the word "acid" in the carboxylic acid is replaced by the word "anhydride."

R-C-O-C-R
  ‖   ‖
  O   O

Acid anhy-
dride

$CH_3CH_2COCCH_2CH_3$

Propionic anhy-
dride

$(ClCH_2CH_2CH_2CO)_2O$

4-Chlorobutanoic
anhydride

Benzoic anhydride

$CH_3CH_2C-O-CCH_3$

Acetic propanoic
anhydride
(a mixed anhydride)

PREPARATION OF ACID ANHYDRIDES

A) Carboxylic acid anhydrides are derived from carboxylic acids by the loss of water.

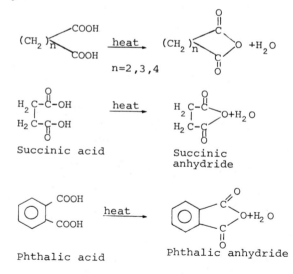

Succinic acid → Succinic anhydride

Phthalic acid → Phthalic anhydride

B) Acid chlorides react with salts of organic acids to yield anhydrides.

$$R-\underset{\underset{O}{\|}}{C}-Cl + R'-\underset{\underset{O}{\|}}{C}-O^{\ominus}Na^{\oplus} \rightarrow R-\underset{\underset{O}{\|}}{C}-O-\underset{\underset{O}{\|}}{C}-R' + NaCl$$

acid   carboxylic acid acid anhydride
chloride   salt

Ex.

$$CH_3COCl + CH_3COONa \rightarrow (CH_3CO)_2O + NaCl$$

acetyl   sodium   acetic anhydride
chloride acetate

C) Ketene, $CH_2 = C = O$, reacts with acid to yield anhydrides.

Ex.

$$CH_3COCH_3 \xrightarrow{700°-750°} CH_2 = C = O + CH_4$$

acetone   ketene

$$CH_2{=}C{=}O + CH_3CH_2COOH \rightarrow CH_3\underset{\underset{O}{\|}}{C}H_2CO\overset{\overset{O}{\|}}{C}CH_3$$

     propanoic acid acetic propionic
            anhydride

$$CH_3COOH \xrightarrow{\frac{AlPO_4}{700°}} H_2O + CH_2 = C = O$$

acetic acid   ketene

$$CH_2 = C = O + CH_3COOH \rightarrow (CH_3CO)_2O$$

       acetic acid acetic anhydride

D) Thallium salts of carboxylic acids react with $SOCl_2$ to give symmetric anhydrides.

$$2RCOOTl + SOCl_2 \quad [(RCOO)_2SO] + 2TlCl \longrightarrow$$

$$(RCO)_2O + SO_2$$

thallium (I)   acid anhydride
 salt

REACTIONS OF ACID ANHYDRIDES

A) Conversion to acids and acid anhydrides

$$(RCO)_2O + HZ \rightarrow RCOZ + RCOOH$$

a) Conversion into acids. Hydrolysis

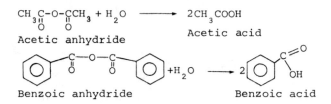

$$CH_3\overset{O}{\underset{\|}{C}}-O-\overset{O}{\underset{\|}{C}}CH_3 + H_2O \longrightarrow 2CH_3COOH$$

Acetic anhydride

Acetic acid

Benzoic anhydride $+H_2O \longrightarrow 2$ Benzoic acid

b) Conversion into amides. Ammonolysis

Ex.

$$(CH_3CO)_2O + 2NH_3 \rightarrow CH_3\overset{}{\underset{\underset{O}{\|}}{C}}-NH_2 + CH_3COO^-\overset{+}{N}H_4$$

acetic anhydride   acetamide   ammonium
                                        acetate

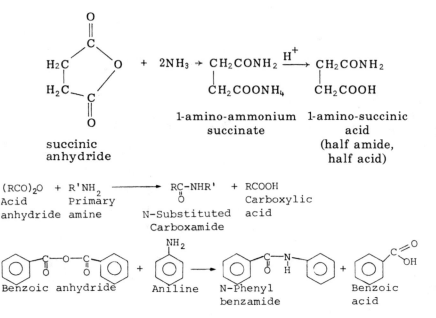

$$\begin{matrix} & O \\ & \| \\ & C \\ H_2C & \diagdown \\ | & \quad O \\ H_2C & \diagup \\ & C \\ & \| \\ & O \end{matrix} \quad + \quad 2NH_3 \rightarrow \begin{matrix} CH_2CONH_2 \\ | \\ CH_2COONH_4 \end{matrix} \overset{H^+}{\longrightarrow} \begin{matrix} CH_2CONH_2 \\ | \\ CH_2COOH \end{matrix}$$

succinic
anhydride

1-amino-ammonium
succinate

1-amino-succinic
acid
(half amide,
half acid)

$$(RCO)_2O + R'NH_2 \longrightarrow R\overset{}{\underset{\underset{O}{\|}}{C}}-NHR' + RCOOH$$
Acid   Primary   Carboxylic
anhydride   amine   N-Substituted   acid
                              Carboxamide

Benzoic anhydride + Aniline $\longrightarrow$ N-Phenyl benzamide + Benzoic acid

c) Conversion into esters. Alcoholysis

$$(CH_3CO)_2O + CH_3OH \longrightarrow CH_3COOCH_3 + CH_3COOH$$

| Acetic anhydride | Methyl alcohol | Methyl acetate (An ester) | Acetic acid |

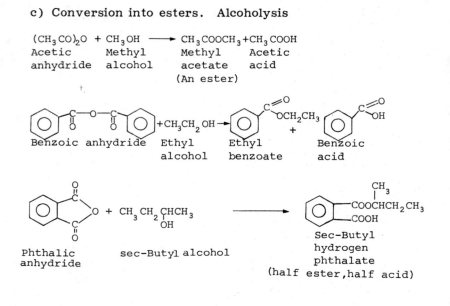

Benzoic anhydride   Ethyl alcohol   Ethyl benzoate   +   Benzoic acid

Phthalic anhydride   +   sec-Butyl alcohol $\longrightarrow$ Sec-Butyl hydrogen phthalate
(half ester, half acid)

B) Reduction to Alcohols

Ex.

$$(CH_3CO)_2O + LiAlH_4 \rightarrow LiAlO(OCH_2CH_3)_2 \xrightarrow{\; H^+ \;} 2CH_3CH_2OH$$

acetic anhydride   ethyl alcohol

C) Formation of Ketones. Friedel-Crafts Acylation.

$$(RCO)_2O + ArH \xrightarrow[\substack{\text{or other} \\ \text{Lewis acid}}]{AlCl_3} R-\underset{\underset{O}{||}}{C}-Ar + RCOOH$$
                                                      A ketone

$$(CH_3CO)_2O + \text{(Mesitylene)} \xrightarrow{AlCl_3} CH_3-\underset{\underset{O}{||}}{C}-\text{(ring)} + CH_3COOH$$

Acetic anhydride   Mesitylene   Methyl mesityl ketone   Acetic acid

106

13.3 ESTERS

NOMENCLATURE (IUPAC SYSTEM)

When naming esters, the alcohol or phenol group is named first, followed by the name of the acid with the "-ic" ending replaced by "-ate." Esters of cycloalkane carboxylic acids have the ending "carboxylate."

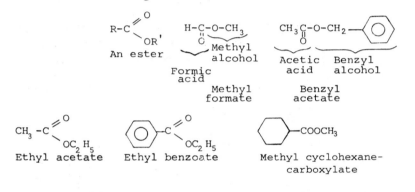

PROPERTIES

The low molecular weight esters are insoluble in water. They are excellent solvents for many organic compounds. Volatile esters have pleasant odors, and they are used in the preparation of perfumes and artificial flavorings.

PREPARATION OF ESTERS

A) From Acids

$$RCOOH + R'OH \underset{}{\overset{H^+}{\rightleftharpoons}} RCOOR' + H_2O$$

Reactivity of R'OH

$1° > 2° > 3°$

Carbox- Alcohol   Ester
ylic   R'is usually
R may   alkyl
be alkyl
or aryl

CH₃COOH + HOH₂C—⟨○⟩ ⇌ CH₃COOCH₂——⟨○⟩+H₂O

Acetic acid   Benzyl alcohol   Benzyl acetate

⟨○⟩—COOH + HOCH₂CH(CH₃)CH₃ ⇌ ⟨○⟩—COOCH₂CHCH₃(CH₃) + H₂O

Benzoic acid   Isobutyl alcohol   Isobutyl benzoate

107

B) From Acid Chlorides

$$\underset{O}{R-\overset{\parallel}{C}-Cl} + \underset{(or\ ArOH)}{R'OH} \longrightarrow \underset{\substack{O \\ (or\ RCOOAr)}}{R-\overset{\parallel}{C}-O-R'} + HCl$$

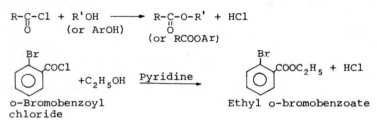

o-Bromobenzoyl chloride $\quad$ Pyridine $\quad$ Ethyl o-bromobenzoate

C) From Anhydrides

$$(RCO)_2O + R'OH(or\ ArOH) \longrightarrow RCOOR'(or\ RCOOAr) + RCOOH$$

$$\underset{\substack{\ \ O\ \ \ \ O}}{CH_3\overset{\parallel}{C}-O-\overset{\parallel}{C}CH_3} + CH_3-OH \longrightarrow \underset{O}{CH_3\overset{\parallel}{C}-OCH_3} + CH_3COOH$$

Acetic anhydride $\qquad$ Methyl alcohol $\qquad$ Methyl acetate

$$(CH_3CO)_2O + HO-\!\!\left\langle\bigcirc\right\rangle\!\!-NO_2 \xrightarrow{NaOH} CH_3COO-\!\!\left\langle\bigcirc\right\rangle\!\!-NO_2 + CH_3COOH$$

Acetic anhydride $\qquad$ p-Nitrophenol $\qquad$ p-Nitrophenyl acetate

D) From salts of very active or heavy metals and alkyl halides.

$$R-COOM + X-R' \rightarrow R-CO-O-R' + MX$$

Ex.

$$CH_3-CO-OAg + I-CH_2CH_3 \rightarrow CH_3-CO-O-CH_2CH_3 + AgI$$
$$\text{ethyl acetate}$$

E) From esters. Transesterification

$$RCOOR' + R''OH \underset{}{\overset{H^+\ or\ OR^-}{\rightleftharpoons}} RCOOR'' + R'OH$$

F) From alcohols and ketene.

$$CH_2{=}C{=}O + ROH \rightarrow CH_3COOR$$
$$\text{ketene} \qquad\qquad \text{ester}$$

G) Methyl esters from diazomethane.

$$RCOOH + CH_2N_2 \rightarrow RCOOCH_3 + N_2$$
methyl ester

H) Preparation of lactones (cyclic esters) and lactides (dilactones). Hydroxy acids undergo self-esterification under acid catalysis.

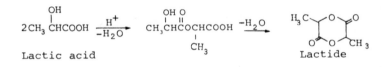

$$HOCH_2CH_2CH_2COOH \underset{\text{}}{\overset{H^+}{\rightleftharpoons}}$$
γ-Hydroxybutyric acid

+H_2O

γ-Butyrolactone

2$CH_3\overset{OH}{\underset{}{C}}HCOOH \xrightarrow[-H_2O]{H^+}$ $CH_3\overset{OH}{\underset{}{C}}H\overset{O}{\underset{}{C}}OCHCOOH \xrightarrow{-H_2O}$

Lactic acid

CH_3

Lactide

REACTIONS OF ESTERS

A) Conversion into acids and acid derivatives.

a) Acid-catalyzed hydrolysis of esters

$$R-\overset{}{\underset{\overset{\|}{O}}{C}}-OR' + H_2O \text{ (excess)} \xrightarrow{H^+} R-\overset{}{\underset{\overset{\|}{O}}{C}}-OH + R'-OH$$

⟨O⟩—$COOC_2H_5$ +H_2O $\xrightarrow{H_2SO_4}$ ⟨O⟩—COOH +C_2H_5OH

Ethyl benzoate

Benzoic acid   Ethyl alcohol

b) Basic hydrolysis. Saponification of esters.

$$R-\overset{}{\underset{\overset{\|}{O}}{C}}-OR' + H_2O \xrightarrow{OH^-} R-\overset{}{\underset{\overset{\|}{O}}{C}}-O^- + R'-OH$$

$$\xrightarrow{H^+} RCOOH$$

⟨O⟩—$COOC_2H_5$ +H_2O $\xrightarrow{NaOH}$ ⟨O⟩—$COO^{\ominus}Na^{\oplus}$ +C_2H_5OH

Ethyl benzoate

Sodium benzoate Ethyl alcohol

$CH_3CH_2COOCH_3$ +H_2O $\xrightarrow{NaOH}$ CH_3CH_2COONa +CH_3OH

109

c) Conversion into amides. Ammonolysis.

$$R-\overset{\overset{\displaystyle O}{\|}}{C}-OR' + NH_3 \rightarrow R-\overset{\overset{\displaystyle O}{\|}}{C}-NH_2 + R'-OH$$

Ex.

$$CH_3COOC_2H_5 + NH_3 \rightarrow CH_3CONH_2 + C_2H_5OH$$

ethyl acetate   acetamide   ethyl alcohol

d) Conversion into esters. Transesterification. Alcoholysis.

$$R-\overset{\overset{\displaystyle O}{\|}}{C}-OR' + R''-OH \underset{\xrightarrow{\hspace{1cm}}}{\overset{\text{acid or}}{\overset{\text{base}}{\rightleftharpoons}}} R-\overset{\overset{\displaystyle O}{\|}}{C}-OR'' + R'-OH$$

Ex.

$$
\begin{array}{l}
CH_2-O-\overset{\overset{\displaystyle O}{\|}}{C}-R \\
CH-O-\overset{\overset{\displaystyle O}{\|}}{C}-R' \\
CH_2-O-\overset{\overset{\displaystyle O}{\|}}{C}-R''
\end{array}
+ CH_3OH \xrightarrow[\text{base}]{\text{acid or}}
\begin{array}{l}
RCOOCH_3 \\
+ \\
R'COOCH_3 \\
+ \\
R''COOCH_3
\end{array}
+
\begin{array}{l}
CH_2OH \\
CHOH \\
CH_2OH
\end{array}
$$

a glyceride   Mixture of   glycerol
(a fat)   methyl esters

B) Reaction with Grignard reagents

$$RCOOR' + 2R''MgX \rightarrow R-\overset{\overset{\displaystyle R''}{|}}{\underset{\underset{\displaystyle OH}{|}}{C}}--R''$$

tertiary alcohol

Ex.

$$CH_3-\overset{\overset{\displaystyle CH_3}{|}}{C}HCOOC_2H_5 + 2CH_3MgI \rightarrow CH_3CH-\overset{\overset{\displaystyle CH_3}{|}}{\underset{\underset{\displaystyle OH}{|}}{C}}-CH_3 \quad (CH_3 \; CH_3)$$

ethyl isobutyrate   methyl mag-   2,3-dimethyl-2-
                    nesium iodide   butanol
                    2 moles

C) Reduction to alcohols

a) Catalytic hydrogenation. Hydrogenolysis.

$$RCOOR' + 2H_2 \xrightarrow[\substack{3000-6000 \\ lb/in^2}]{\substack{CuO, CuCr_2O_4 \\ 250°C}} RCH_2OH + R'OH$$

primary alcohol

Ex.

$$CH_3-\underset{\underset{CH_3}{|}}{\overset{\overset{CH_3}{|}}{C}}-COOC_2H_5 + 2H_2 \xrightarrow[250°, 3300 \ lb/in^2]{CuO, CuCr_2O_4} $$

ethyl trimethyl
acetate (ethyl
2,2-dimethyl
propanoate)

$$CH_3-\underset{\underset{CH_3}{|}}{\overset{\overset{CH_3}{|}}{C}}-CH_2OH + C_2H_5OH$$

| Neopentyl alcohol (2,2-dimethyl propanol) | ethyl alcohol |

b) Bouvaeult-Blance method.

$$RCOOR' \xrightarrow{Na + alcohol} RCH_2OH + R'OH$$

c) Lithium aluminum hydride reduction.

$$4RCOOR' + 2LiAlH_4 \xrightarrow[ether]{anhyd} \begin{matrix} LiAl(OCH_2R)_4 \xrightarrow{H^+} \\ LiAl^+(OR')_4 \end{matrix} \begin{matrix} RCH_2OH \\ + \\ R'OH \end{matrix}$$

Ex.

$$CH_3(CH_2)_7CH = CH(CH_2)_7COOCH_3 \xrightarrow{LiAlH_4}$$

methyl oleate
(methyl cis-9-
octadecenoate)

$$CH_3(CH_2)_7CH = CH(CH_2)_7CH_2OH$$

oleyl alcohol (cis-9-octadecene
-1-ol)

D) Reaction with carbanions. Claisen condensation.

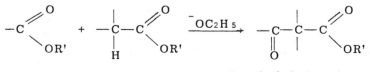

A β-keto ester

111

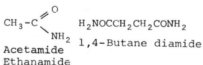

C_6H_5—COOC$_2$H$_5$ +CH$_3$COOC$_2$H$_5$ $\xrightarrow{^-OC_2H_5}$ C_6H_5—$\overset{O}{\underset{||}{C}}$—CH$_2$COOC$_2H_5$

Ethyl benzoate   Ethyl acetate   Ethyl benzoyl
                                                            acetate
                                                            +C$_2$H$_5$OH

C_2H_5O—$\overset{O}{\underset{||}{C}}$—OC$_2H_5$ +C$_6$H$_5$CH$_2$COOC$_2$H$_5$ $\xrightarrow{C_2H_5O^{\ominus}}$

Ethyl carbon-   Ethyl phenylacetate
ate
                     $\overset{C_6H_5}{|}$
            C_2H_5O—$\overset{O}{\underset{||}{C}}$—CHCOOC$_2H_5$ +C$_2$H$_5$OH

            Ethyl phenylmalonate
            Phenylmalonic ester

2CH$_3$$\overset{O}{\underset{||}{C}}OCH_2CH_3$ $\xrightarrow{C_2H_5O^{\ominus}}$ CH$_3$$\overset{O}{\underset{||}{C}}CH_2$COOCH$_2CH_3$

Ethyl acetate   Ethyl acetoacetate
2 moles   Aceto acetic ester

13.4 AMIDES

NOMENCLATURE (IUPAC SYSTEM)

When naming amides, the "-ic acid" of the common name (or the "-oic acid" of the IUPAC name) of the parent acid is replaced by "amide." Amides of cycloalkane carboxylic acids have the ending carboxamide.

CH$_3$—C$\overset{\diagup O}{\diagdown NH_2}$ H$_2$NOCCH$_2$CH$_2$CONH$_2$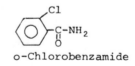

Acetamide   1,4-Butane diamide   o-Chlorobenzamide
Ethanamide

$\square$—CONH$_2$ CH$_3$$\overset{O}{\underset{||}{C}}$—NH—CH$_2CH_3$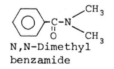

Cyclobutane-   N-Ethyl acet-   N,N-Dimethyl
carboxamide   amide   benzamide

PROPERTIES OF AMIDES

Amides have high boiling points due to strong intermolecular hydrogen bonding. Amides with up to five or

112

six carbons are soluble in water. Like the other acid derivatives, they are soluble in the usual organic solvents. Primary amides are very weak basic compounds and are insoluble in dilute acids.

PREPARATION OF AMIDES

A) Amides from acid chlorides and ammonia or amines.

$$RCOCl + 2NH_3 \rightarrow RCONH_2 + NH_4Cl$$

an amide (1°)

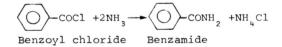

Benzoyl chloride   Benzamide

$$RCOCl + R'NH_2 \rightarrow RCONHR' + HCl$$

1° amine   2° amide

$$RCOCl + 2NHR'_2 \rightarrow RCONR'_2 + R'^{\oplus}_2 NH_2Cl^-$$

2° amine   3° amide

B) From acid anhydrides and ammonia or amines.

$$(RCO)_2O + 2NH_3 \rightarrow RCONH_2 + RCOONH_4$$

amide (1°)

Ex.

$$(CH_3CO)_2O + 2NH_3 \rightarrow CH_3CONH_2 + CH_3COO^-NH_4^+$$

acetic anhydride   acetamide   ammonium acetate

$$(RCO)_2O + 2R'NH_2 \rightarrow RCONHR' + R'NH_3^+RCO_2^-$$

1° amine   2° amide

$$(RCO)_2O + NHR'_2 \rightarrow RCONR'_2 + RCOOH$$

2° amine   3° amide

C) Amides from esters by ammonolysis.

$$R\overset{\text{O}}{\underset{\|}{-C}}-O-R' + NH_3 \rightarrow RCONH_2 + R'OH$$

1° amide

D) Amides by the pyrolysis of ammonium carboxylates.

$$\underset{\underset{O}{\|}}{R-C-OH} + NH_3 \rightarrow \underset{\underset{O}{\|}}{R-C-O^{\ominus}NH_4^{\oplus}} \xrightarrow{heat} \underset{\underset{O}{\|}}{R-C-NH_2} + H_2O$$

ammonium   1° amide
carboxylate

E) From nitriles.

$$RC \equiv N + H_2O \xrightarrow[\text{or } OH^-]{H^+} RCONH_2$$

1° amide

Ex.

$$CH_3CH_2C \equiv N + H_2O \xrightarrow[\text{conc.}]{H_2SO_4} CH_3CH_2CONH_2$$

propionamide

F) Amides from amines and ketene.

$$RNH_2 + CH_2 = C = O \rightarrow CH_3CONHR$$

2° amide

G) Beckmann rearrangement

$$\underset{\underset{O}{\|}}{:N-OH} \xrightarrow{H^+} \underset{}{:N-^+OH_2} \longrightarrow R-N=C^+-R' \xrightarrow{H_2O}$$
$$R-CR' \qquad R-C-R'$$

R group trans
to OH migrates

$$\underset{\underset{^+OH_2}{|}}{R-N=C-R'} \xrightarrow[-H^+]{} \underset{\underset{H-O}{|}}{R-N=C-R'} \longleftarrow \rightleftharpoons \underset{\underset{H \quad O}{| \quad \|}}{R-N-C-R'}$$

enol form of   more stable keto
amide   form of amide

H) Preparation of secondary and tertiary amides.

$$RCONH_2 + (CH_3CO)_2O \xrightarrow{heat} RCONHCOCH_3 \xrightarrow[\text{heat}]{(CH_3CO)_2O}$$

2° amide

$$RCON(COCH_3)_2$$

3° amide

I) Preparation of lactams (cyclic amides).

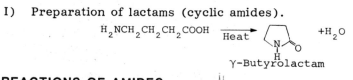

γ-Butyrolactam

REACTIONS OF AMIDES

A) Hydrolysis

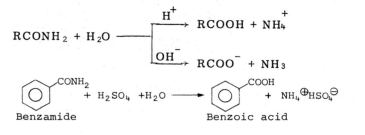

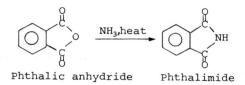

Benzamide   Benzoic acid

$$CH_3CH_2CH_2CONH_2 + NaOH + H_2O \longrightarrow CH_3CH_2CH_2COO^{\ominus}Na^{\oplus} + NH_3$$
Butyramide   Sodium butyrate

B) Conversion into imides.

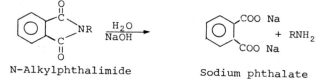

Phthalic anhydride   Phthalimide

C) Salt Formation. Gabriel's synthesis.

Primary amides form salts which are hydrolyzed by the stronger base water. The hydrogen of secondary amides is more acidic and the stable salts are formed with aqueous NaOH.

N-Alkylphthalimide   Sodium phthalate

D) Reaction with nitrous acid.

$$RCONH_2 + ONOH \rightarrow RCOOH + N_2 + H_2O$$

E) Dehydration of primary amine into nitrile by treatment with a strong dehydrating agent.

$$RCONH_2 + P_2O_5 \rightarrow RC \equiv N + 2HPO_3$$

F) Reduction of amides by lithium aluminum hydride.

$$R-\underset{\underset{O}{\|}}{C}-NH_2 \xrightarrow{LiAlH_4} \xrightarrow{H_2O} RCH_2-NH_2$$

primary amide   primary amine

$$R-\underset{\underset{O}{\|}}{C}-NH-R' \xrightarrow{LiAlH_4} \xrightarrow{H_2O} R-CH_2-NH-R'$$

N-substituted amide   secondary amine
secondary amide

$$R-\underset{\underset{O}{\|}}{C}-NR'_2 \xrightarrow{LiAlH_4} \xrightarrow{H_2O} R-CH_2-N-R'_2$$

N,N-disubstituted   tertiary amine
amide
tertiary amide

G) Hoffman degradation of amides. Hypobromite reaction

$$RCONH_2 + NaOBr + 2NaOH \rightarrow RNH_2 + Na_2CO_3 + NaBr + H_2O$$

1° amide   1° amine

Ex.

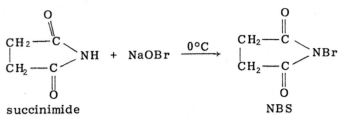

succinimide   3-Aminopropanoic acid

H) N-Bromosuccinimide (NBS)

succinimide   NBS

NBS is a very useful reagent for the bromination of methyl or methylene groups adjacent to double or triple bonds.

THE PERIODIC TABLE

METALS

NONMETALS

TRANSITION METALS

KEY

Atomic weight →	112.40
	Cd
Symbol / Atomic number →	48

PERIODS	IA	IIA	IIIB	IVB	VB	VIB	VIIB	VIII	VIII	VIII	IB	IIB	IIIA	IVA	VA	VIA	VIIA	O
1	1.0079 **H** 1																	4.00260 **He** 2
2	6.94 **Li** 3	9.01218 **Be** 4											10.81 **B** 5	12.011 **C** 6	14.0067 **N** 7	15.9994 **O** 8	18.9984 **F** 9	20.179 **Ne** 10
3	22.9898 **Na** 11	24.305 **Mg** 12											26.9815 **Al** 13	28.086 **Si** 14	30.9738 **P** 15	32.06 **S** 16	35.453 **Cl** 17	39.948 **Ar** 18
4	39.098 **K** 19	40.08 **Ca** 20	44.9559 **Sc** 21	47.90 **Ti** 22	50.9414 **V** 23	51.996 **Cr** 24	54.9380 **Mn** 25	55.847 **Fe** 26	58.9332 **Co** 27	58.71 **Ni** 28	63.546 **Cu** 29	65.38 **Zn** 30	69.72 **Ga** 31	72.59 **Ge** 32	74.9216 **As** 33	78.96 **Se** 34	79.904 **Br** 35	83.80 **Kr** 36
5	85.4678 **Rb** 37	87.62 **Sr** 38	88.9059 **Y** 39	91.22 **Zr** 40	92.9064 **Nb** 41	95.94 **Mo** 42	98.9062 **Tc** 43	101.07 **Ru** 44	102.9055 **Rh** 45	106.4 **Pd** 46	107.868 **Ag** 47	112.40 **Cd** 48	114.82 **In** 49	118.69 **Sn** 50	121.75 **Sb** 51	127.60 **Te** 52	126.9046 **I** 53	131.30 **Xe** 54
6	132.9054 **Cs** 55	137.34 **Ba** 56	57–71 *	178.49 **Hf** 72	180.9479 **Ta** 73	183.85 **W** 74	186.2 **Re** 75	190.2 **Os** 76	192.22 **Ir** 77	195.09 **Pt** 78	196.9665 **Au** 79	200.59 **Hg** 80	204.37 **Tl** 81	207.2 **Pb** 82	208.9804 **Bi** 83	(210) **Po** 84	(210) **At** 85	(222) **Rn** 86
7	(223) **Fr** 87	(226.0254) **Ra** 88	89–103 †	(260) **Ku** 104	(260) **Ha** 105													

* LANTHANIDE SERIES	138.9055 **La** 57	140.12 **Ce** 58	140.9077 **Pr** 59	144.24 **Nd** 60	(145) **Pm** 61	150.4 **Sm** 62	151.96 **Eu** 63	157.25 **Gd** 64	158.9254 **Tb** 65	162.50 **Dy** 66	164.9304 **Ho** 67	167.26 **Er** 68	168.9342 **Tm** 69	173.04 **Yb** 70	174.97 **Lu** 71
† ACTINIDE SERIES	(227) **Ac** 89	232.0381 **Th** 90	231.0359 **Pa** 91	238.029 **U** 92	237.0482 **Np** 93	(242) **Pu** 94	(243) **Am** 95	(245) **Cm** 96	(245) **Bk** 97	(248) **Cf** 98	(253) **Es** 99	(254) **Fm** 100	(256) **Md** 101	(253) **No** 102	(257) **Lr** 103

REA's Problem Solvers

The "PROBLEM SOLVERS" are comprehensive supplemental text-books designed to save time in finding solutions to problems. Each "PROBLEM SOLVER" is the first of its kind ever produced in its field. It is the product of a massive effort to illustrate almost any imaginable problem in exceptional depth, detail, and clarity. Each problem is worked out in detail with a step-by-step solution, and the problems are arranged in order of complexity from elementary to advanced. Each book is fully indexed for locating problems rapidly.

ACCOUNTING
ADVANCED CALCULUS
ALGEBRA & TRIGONOMETRY
AUTOMATIC CONTROL
  SYSTEMS/ROBOTICS
BIOLOGY
BUSINESS, ACCOUNTING, & FINANCE
CALCULUS
CHEMISTRY
COMPLEX VARIABLES
COMPUTER SCIENCE
DIFFERENTIAL EQUATIONS
ECONOMICS
ELECTRICAL MACHINES
ELECTRIC CIRCUITS
ELECTROMAGNETICS
ELECTRONIC COMMUNICATIONS
ELECTRONICS
FINITE & DISCRETE MATH
FLUID MECHANICS/DYNAMICS
GENETICS
GEOMETRY

HEAT TRANSFER
LINEAR ALGEBRA
MACHINE DESIGN
MATHEMATICS for ENGINEERS
MECHANICS
NUMERICAL ANALYSIS
OPERATIONS RESEARCH
OPTICS
ORGANIC CHEMISTRY
PHYSICAL CHEMISTRY
PHYSICS
PRE-CALCULUS
PROBABILITY
PSYCHOLOGY
STATISTICS
STRENGTH OF MATERIALS &
  MECHANICS OF SOLIDS
TECHNICAL DESIGN GRAPHICS
THERMODYNAMICS
TOPOLOGY
TRANSPORT PHENOMENA
VECTOR ANALYSIS

*If you would like more information about any of these books,
complete the coupon below and return it to us or visit your local bookstore.*

RESEARCH & EDUCATION ASSOCIATION
61 Ethel Road W. • Piscataway, New Jersey 08854
Phone: (908) 819-8880

Please send me more information about your Problem Solver Books

Name _____

Address _____

City _____ State _____ Zip _____